FORSCHUNGSBERICHTE AUS DEM LEHRSTUHL FÜR REGELUNGSSYSTEME

TECHNISCHE UNIVERSITÄT KAISERSLAUTERN

Band 20

Forschungsberichte aus dem Lehrstuhl für Regelungssysteme

Technische Universität Kaiserslautern

Band 20

Herausgeber:

Prof. Dr. Steven Liu

Zhuoqi Zeng

Ultra-wideband Based Indoor Localization Using Sensor Fusion and Support Vector Machine

Logos Verlag Berlin

Forschungsberichte aus dem Lehrstuhl für Regelungssysteme
Technische Universität Kaiserslautern

Herausgegeben von
Univ.-Prof. Dr.-Ing. Steven Liu
Lehrstuhl für Regelungssysteme
Technische Universität Kaiserslautern
Erwin-Schrödinger-Str. 12/332
D-67663 Kaiserslautern
E-Mail: sliu@eit.uni-kl.de

Bibliographic information published by the Deutsche Nationalbibliothek

The Deutsche Nationalbibliothek lists this publication in the Deutsche Nationalbibliografie; detailed bibliographic data are available on the Internet at http://dnb.d-nb.de .

ISBN 978-3-8325-5229-9
ISSN 2190-7897

Logos Verlag Berlin GmbH
Georg-Knorr-Str. 4, Geb. 10,
12681 Berlin
Tel.: +49 (0)30 / 42 85 10 90
Fax: +49 (0)30 / 42 85 10 92
http://www.logos-verlag.de

Ultra-wideband Based Indoor Localization Using Sensor Fusion and Support Vector Machine

Ultra-Breitband basierte Indoor-Lokalisierung mit Sensordatenfusion und Support Vector Machine

Vom Fachbereich Elektrotechnik und Informationstechnik
der Technischen Universität Kaiserslautern
zur Verleihung des akademischen Grades
Doktor der Ingenieurwissenschaften (Dr.-Ing.)
genehmigte Dissertation

von

M. Sc. Zhuoqi Zeng
geboren in Hunan, China

Acknowledgment

This thesis presents the results of my work at two locations: firstly at Bosch (China) Investment Ltd., Research and Technology Center Asia/Pacific (CR/RTC5-AP), in Shanghai; secondly at the Institute of Control Systems (LRS), Department of Electrical and Computer Engineering, at the University of Kaiserslautern.

First and foremost, I would like to express my great gratitude to my supervisor Prof. Dr. -Ing. Steven Liu, the head of the Institute of Control Systems at the University of Kaiserslautern, for the excellent supervision of my research work and the valuable scientific discussion. He has been a great mentor and gives me a lot of valuable advice during my work. I would also like to thank my supervisor Prof. Dr. Lei Wang, at College of Electronics and Information Engineering, TongJi University and my supervisor from the company side: Dr. Rose Hannes, Dr. Yu Huajun, Dr. Wang William at Bosch (China) Investment Ltd. Thanks also go to Prof. Dr. Ping Zhang, the head of the Institute of Automatic Control at the University of Kaiserslautern, for joining the thesis committee.

My time at Bosch (China) Investment Ltd. and the Institute of Control Systems has been very enjoyable and rewarding. I am grateful to all the colleagues for the open discussions and great supports.

Finally, I would like to thank my parents Lingfang Zuo, Jianjun Zeng, my uncle Yaoshu Peng, my aunt Jingyu Yang, my best friend Guojun Hou and his lovely wife Lihua Zhang for the love and great support throughout all these years.

Kaiserslautern, Octorber 06, 2020
Zhuoqi Zeng

Abstract

To improve the productivity and flexibility of industrial robots in conventional production lines, Bosch (China) Investment Ltd. has proposed a project called "Real-time Safety Virtual Positioning" (RSVP) to replace the fixed installation of safeguards with an indoor localization based safety system. After the comparison of the existing indoor localization systems, the Ultra-wideband (UWB) system is deemed suitable for the project based on the following perspectives: localization accuracy, coverage area, complexity and cost. The greatest challenge in guaranteeing the UWB position estimation accuracy is identifying the Non-line-of-sight (NLOS) measurements. An updated overview and comparison of the current NLOS identification methods are missing.

Hence, this thesis presents an updated overview of these approaches. After a comparison, the channel impulse response (CIR) and IMU sensor fusion based approaches are proved to be the most effective NLOS detection methods. However, there is a lack of optimization of the feature combination and parameters in machine learning algorithms for the current CIR based approaches and these approaches only work for time of arrival (TOA). Furthermore, few studies have compared the localization performance of TOA and time difference of arrival (TDOA) in different environments. The extended Kalman filter based UWB/inertial measurement unit (IMU) fusion algorithm detects the NLOS outliers under the assumption that the errors are Gaussian distributed. However, the NLOS errors are usually not Gaussian distributed, which means that improper LOS/NLOS error models are assumed and used for identification due to the lack of the investigation for measurement errors.

To further improve the NLOS detection and mitigation performance, this thesis systematically investigates the UWB LOS/NLOS errors. The LOS errors are evaluated in different environments and with different distances. Different blockage materials and blockage conditions are considered for NLOS errors. The UWB signal propagation is also investigated. Furthermore, the relationships between the CIRs and the accurate/inaccurate range measurements are theoretically discussed in three different situations: ideal LOS path, small-scale fading: multipath and NLOS path. These theoretical relationships are validated with real measured CIRs in the Bosch Shanghai office environment.

Based on the error and signal propagation investigation results, four different algorithms are proposed for four different scenarios to improve the NLOS identification accuracy. After the comparison of the localization performance for TOA/TDOA, it is found that on normal office floor, TOA works better than TDOA. In harsh industrial environments, where NLOS frequently occurs, TDOA is more suitable than TOA. Thus, in the first scenario, the position estimation is realized with TOA on the office floor, while in the second scenario, a novel approach to combined TOA and TDOA with accurate range and range difference selection is proposed in the harsh industrial environment. The optimization of the feature combination and parameters in machine learning algorithms for accurate measurement detection is discussed for both scenarios. For the third and fourth scenarios, the UWB/IMU fusion system stays in focus. Instead of detecting the NLOS outliers by assuming that the error distributions are Gaussian, the accurate measurement detection is realized based on the triangle inequality theorem. All the proposed approaches are tested with the collected measurements from the developed UWB system. The position estimation of these approaches has better accuracy than that of the traditional methods.

Zusammenfassung

Um die Produktivität und Flexibilität von Industrierobotern in hochautomatisierten Produktionslinien zu verbessern, wurde von Bosch (China) Investment Ltd. ein Projekt namens "Real-time Safety Virtual Positioning" (RSVP) ins Leben gerufen, das die feste Installation von Schutzvorrichtungen durch ein auf Indoor-Lokalisierung basierendes Sicherheitssystem ersetzt. Auf Basis eines Vergleichs von bestehenden Indoor-Lokalisierungssystemen wurde das Ultra-Breitbandsystem (UWB) als für das Projekt geeignetes System aus folgenden Gründen ausgewählt: Lokalisierungsgenauigkeit, Abdeckungsbereich, Komplexität und Kosten. Die größte Herausforderung bei der Gewährleistung der UWB-Positionsschätzgenauigkeit besteht in der Identifikation für Nicht Sichtverbindung (NLOS). Eine aktualisierte Übersicht und ein Vergleich der aktuellen NLOS-Identifikationsmethoden fehlen.

In dieser Arbeit wird zunächst eine Überblicke über die vorhandenen NLOS Identifikationsansätze vorgestellt. Nach dem Vergleich dieser Ansätze haben sich die auf Kanalimpulsantworten (CIR)- und inertiale Messeinheit (IMU) Sensorfusion basierenden Ansätze als die effektivsten NLOS Identifikationsansätze herausgestellt. Die Optimierung der Feature Kombination und der Parameter in Maschinelles Lernen Algorithmen für die aktuellen CIR-basierten Ansätze fehlt jedoch, und diese Ansätze funktionieren nur für Time of Arrival (TOA). Der Vergleich der Lokalisierungsqualität von TOA und Time Difference of Arrival (TDOA) in verschiedenen Umgebungen fehlt. Der Extended Kalman Filter basierte UWB/inertiale Messeinheit (IMU) Fusionsalgorithmus erkennt die NLOS-Ausreißer unter der Annahme, dass die Fehler gaußverteilt sind. Meistens sind die NLOS-Fehler jedoch nicht gaußverteilt, was bedeutet, dass aufgrund der fehlenden Messfehleruntersuchung ein falsches LOS/NLOS-Fehlermodell angenommen und zur Identifizierung verwendet wird.

Um die NLOS Erkennungs- und Lokalizationsgenauigkeit weiter zu verbessern, werden die UWB LOS/NLOS Fehler systematisch untersucht. Die LOS-Fehler werden in verschiedenen Umgebungen und mit unterschiedlichen Entfernungen ausgewertet. Unterschiedliche Sperrmaterialien und Sperrbedingungen werden für NLOS-Fehler berücksichtigt. Zudem wird die UWB-Signalausbreitung untersucht. Darüber hinaus werden die Beziehungen zwischen den CIRs und den genauen/ungenauen Entfernungsmessungen in drei verschiedenen Situationen diskutiert: idealer LOS, NLOS und Mehrwegempfang. Diese theoretischen Zusammenhänge werden mit real gemessenen CIRs in Bosch Shanghai Büroumgebungen validiert.

Basierend auf den Untersuchungsergebnissen des Fehlers und der Signalausbreitung werden vier verschiedene Algorithmen für vier unterschiedliche Szenarien entwicklt um die Genauigkeit der NLOS-Identifizierung zu verbessern. Aus dem Vergleich der Lokalisierungsgenauigkeit für TOA/TDOA ergibt sich, dass TOA in Umgebungen mit wenigen NLOS Situationen (z.B. Normale Bürobereich) bessere Ergebnisse liefert als TDOA. In Industrieumgebungen, in denen NLOS häufig auftritt, ist TDOA im Vergleich zu TOA besser geeignet. So wird im ersten Szenario die Positionsschätzung mit TOA in einer Büroetage realisiert, während im zweiten Szenario eine kombinierte TOA und TDOA Methode zur Lokalisierung in industriellen Umgebungen verwendet wird. Die Optimierung der Featurekombination und der Parameter in maschinelles Lernen Algorithmen zur genauen Messdatenidentifizierung wird für beide Szenarien diskutiert. Für das dritte und vierte Szenario steht das UWB/IMU-Fusionssystem im Fokus. Anstatt die NLOS-Ausreißer zu identifizieren unter der Annahme dass die Fehlerverteilungen gaußschen sind, werden die genauen Messdaten auf Basis der Dreiecksungleichung ausgewählt. Alle vorgeschlagenen Ansätze werden mit den gesammelten Messdaten aus dem entwickelten UWB-System getestet. Die Genauigkeit der Positionsschätzung mit diesen Ansätzen ist besser im Vergleich zu den traditionellen Methoden.

Contents

List of Figures

List of Tables

Notation

Acronyms

AML Approximate maximum likelihood method

ANN Artificial neural networks

AOA Angle of arrival

AR Augmented reality

BS Base station

CIR Channel impulse response

EKF Extended Kalman Filter

GPS Global Positioning System

HDR Heuristic heading drift reduction

HNLOS Hard NLOS

IEKF Iterative extended Kalman filter

IR Infrared

KF Kalman filter

Lidar Light detection and ranging

LOS Line-of sight

LSD Long safety distance

LS Least squares method

MS Mobile station

NLOS Non-line-of sight

pdf Probability density function

PDR Pedestrian dead-reckoning

PF Particle filter

RFID Radio-frequency identification

RSS Received signal strength

RSVP Real-time Safety Virtual Positioning

SLAM Simultaneous localization and mapping

SNLOS Soft non-line-of sight

SSD Short safety distance

SVM Support-vector machine

TDOA Time difference of arrival

TOA Time of arrival

TS Taylor series method

UWB Ultra Wideband

ZARU Zero angular rate update

ZUPT Zero-velocity update

1 Introduction

Localization is a very important technology to promote the quality of human life. The awareness of one's own location is essential when we are in a totally new environment. Outdoor localization can be realized with the Global Positioning System (GPS). However, GPS does not work well indoors due to the blockage of buildings. Nowadays more and more automated robots are used in industrial indoor environments and the accurate position estimation for these robots is essential for the robots collaboration. The precondition for human-robot collaboration is that the distance between human and robot can be accurate detected. Thus the position of the workers also need to be accurate tracked. In the above mentioned and lots of other indoor applications, the indoor position estimation plays a crucial role. Many technologies have been developed for indoor localization, such as ultra wideband (UWB), infrared (IR) and light detection and ranging (Lidar). The indoor localization system can be used in many areas, for instance, in healthcare to track patients' position so that their safety can be improved, in the retail industry to improve the supply chain, or in logistics to optimize the overall workflow, etc. For different applications, a suitable localization system can be determined according to cost of the system, the required accuracy, the system capacity, and so forth.

In a conventional production line, industry robots need to be isolated by safeguards (e.g. safety fence, light curtain.) to guarantee the safety of the workers. The main disadvantage of using the fixed installation of safeguards is the reduction of productivity and flexibility. The fast free movement of industrial robots can dramatically improve flexibility, and various tasks can be finished by different combinations of these robots, so that the production line can have a faster response to rapid market-demanded change. Thus, the CR department in Bosch (China) Investment Ltd. has built a project called "Real-time Safety Virtual Positioning" (RSVP) to enable agile production systems by removing the fixed safeguard installation with the help of an safety indoor localization system. Instead of isolating the robots with safeguards, the safety of the workers is guaranteed based on the continuously measured distance between the robots and the workers. Two different circle zones are defined: the warning zone and the danger zone. The radius of the danger zone is the short safety distance (SSD), while the radius of the warning zone is the long safety distance (LSD), as shown in Fig. 1.1. If a human is in the warning zone, the robot moves more slowly. Once the human steps into the danger zone, the robot stops immediately. With the help of this function, the safeguards can be replaced with the indoor localization based safety system [WZD+19].

One of the most important parts of this system is the accurate position estimation of the robots and humans. Other factors, such as update rate, system capacity, system complexity and coverage., also need to be considered. Thus, a survey for the most widely used indoor localization system is conducted. This section presents an overview and comparison of these systems. It also shortly introduces the RSVP project and explains why UWB is the most suitable localization system for this project. Finally, it presents the main contributions and the structure of this thesis.

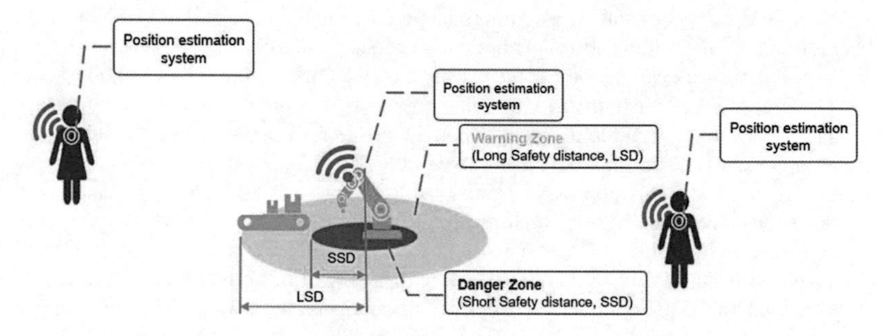

Figure 1.1: Concept of the safety function system in RSVP

1.1 Indoor Localization Systems

GPS is used to realize outdoor localizations. However it does not work well indoors, since the signals can be blocked by buildings. Thus, different systems have been developed for accurate indoor position estimation, such as UWB, IR and Lidar. Based on the utilized technologies and the position estimation algorithms, the most widely used localization systems can be roughly divided into four different groups: wireless based systems, IMU based infrastructure-free indoor localization, SLAM and visible light or acoustic based systems. This is shown in Fig. 1.2 [ZWL19].

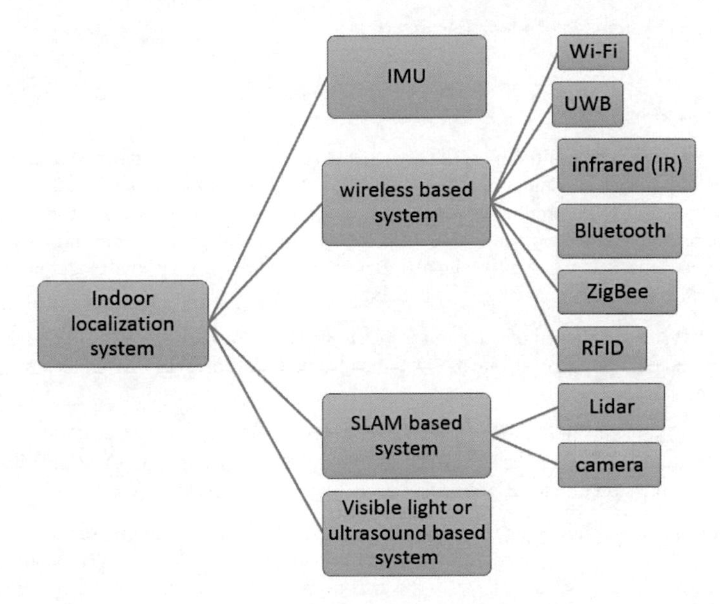

Figure 1.2: Overview of the most widely used indoor localization systems [ZWL19]

1.1.1 Wireless Based Indoor localization

In wireless systems, the information or power is transferred without wires or cables. The basic architecture for a wireless based position estimation system contains three parts: base stations (BSs), mobile stations (MSs) and the software to calculate the position of the MSs based on the measurements. The MSs are used to send the wireless signal to the BSs. Given the received signals in the BSs, different measurements can be obtained, such as the range between a BS and MS, the range difference, the received signal strength or the angle of arrival. Depending on the type of the measurements, different algorithms can be used for position estimation in wireless systems, such as the time of arrival (TOA), time difference of arrival (TDOA), angle of arrival (AOA) and received signal strength (RSS) [LJXG18], [CLL18], [SMS12], [ZYC+13], [GZT08a]. These algorithms are realized in the software so that the real-time position estimation of the MS can be achieved.

Wireless Systems for Indoor Localization

Wi-Fi, UWB, IR, Bluetooth, ZigBee and radio-frequency identification (RFID) are all wireless technologies.

UWB: UWB is one of the most widely used wireless indoor localization systems due to its low-power consumption, high accuracy, robust operation in harsh indoor environments and low complexity for indoor applications. BSs and MSs are the basic components in UWB systems. The MSs send short signal pulses over a broad spectrum, and the BSs receive the signals. Based on the arrival time, the range or range difference measurements can be calculated.

Wi-Fi: Wi-Fi follows the standards in IEEE 802.11. Since Wi-Fi infrastructures already exist, the low cost is one of the main advantages of the Wi-Fi based localization system.

Zigbee: Zigbee technology is based on the standards in IEEE 802.15.4. Three bands can be used in Zigbee: the 2.4 GHz ISM band, 915 MHz band, and 868MHz band. The transmission rate for Zigbee is between 20 kbps and 250 kbps [OCC+17].

Bluetooth: As a short-range wireless technology, Bluetooth is widely used in mobile phones and computers, etc. to transmit information in short ranges. The main advantages of Bluetooth are its lower power consumption, lower costs, and smaller size [HAG17].

RFID: The readers, tags, and servers are the basic components of RFID. The tags are identified and tacked based on the electromagnetic fields in RFID. There are three types of tags: passive , active and semi-passive tags. There are no batteries in the passive tag. The energy used in this tag is obtained from a nearby RFID reader. Active tags have an internal power supply (e.g. battery). Thus, information can be actively sent from the active tag to the reader. Compared to passive tags, active tags are larger and more expensive, but they have more functionalities [Bd08]. Semi-passive tags have internal power supply to power the circuitry, but the way of communication between tags and readers is the same as passive tags.

Infrared: infrared wavelengths are longer than those of visible light. Thus, infrared is invisible to humans. However, humans can feel it as heat. Infrared can be used for localization.

Position Estimation Algorithms for Wireless Systems

The output measurements of these wireless systems can be range, range difference, angle of arrival or received signal strength. Depending on the type of measurement, four different localization algorithms can be used for position estimation of the MS:

TOA, TDOA, AOA and RSS.

1) TOA: The range, which is defined as the distance between the MS and the BS, is used in TOA for position estimation. The position of the tag must lie on the circle that is centered at the BS. In the ideal case, the MS position is the unique intersection point of at least three different circles, with the ranges as the radii. In Fig. 1.3(a), the black intersection point is the position of the MS. The centers of the circles are the positions of the BSs. However, due to the existence of system noise error, the intersection of these circles is not a point but an area. The real MS position can be anywhere in the area. The larger the noise error, the greater is the intersection area and the more inaccurate the obtained position estimation for the MS might be. In Fig. 1.3(b), the real position of the MS can be anywhere in the green intersection area. Many algorithms have been developed to reduce the noise error, such as the least squares method (LS), the Taylor series method (TS), the approximate maximum likelihood method (AML) [GZT08a], and the Kalman filter (KF) [YDH16].

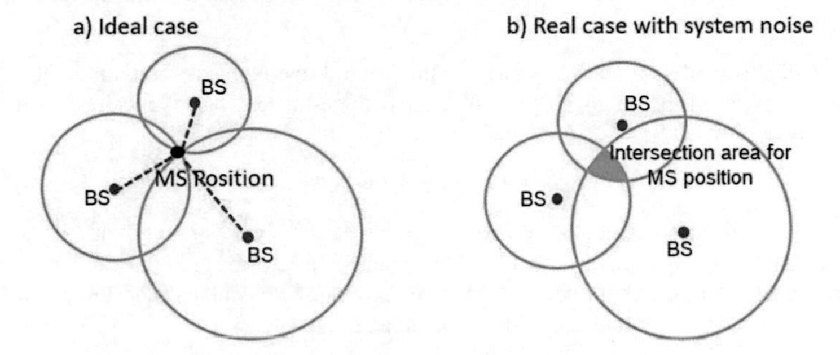

Figure 1.3: (a) TOA localization in an ideal case; (b) TOA localization in a real case with system noise error

2) TDOA: The difference between two ranges can also be used for position estimation. As shown in Fig. 1.4(a), if the difference between the distance from MS to BS1 and the distance from MS to BS2 is constant, then the trajectory of the MS is a hyperbolas. Thus, the intersection of the hyperbola, which are generated based on the range differences with foci at the BSs, is the position of the MS, as shown in Fig. 1.4(a).

3) AOA: Localization can also be achieved with the measured angles. As shown in Fig. 1.4(b), the angles α_1 and α_2 are the measurements. The position of the MS is the black intersection point of two straight lines.

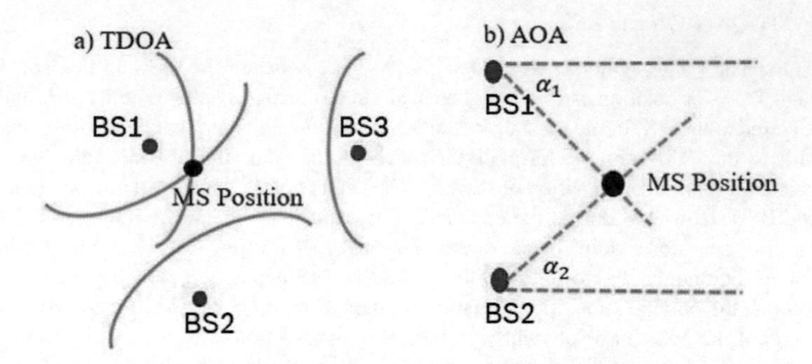

Figure 1.4: (a) TDOA localization; (b) AOA localization

4) RSS: Two different methods can be used to realize the localization based on RSS. The first one is based on the pathloss model, and the second on the fingerprint algorithm.

4.a) Pathloss model based algorithm: In this method, the pathloss model is built and used to determine the range. For example, the pathloss model for UWB can be written as follows [GJD+07]:

$$P_r = P_0 - 10nlog_{10}\frac{d}{d_0} + S \tag{1.1}$$

where d_0 is the reference distance, and P_0 is the received signal strength with distance d_0. The pathloss exponent is represented by n and S is a zero-mean log-normal random variable. Based on this model, the UWB ranges can be calculated and the position estimation can then be realized with the obtained ranges.

4.b) Fingerprint based algorithm: This method comprises two different phases, an off-line and an on-line phase. In the off-line phase, the RSSs at different reference locations are collected to build the database. During the on-line phase, the new measured RSS is compared with the collected RSSs in the database. Based on the comparison, the position estimation can be realized [YXW17].

1.1.2 IMU Based Infrastructure-free Indoor Localization

A 9-axis IMU system contains a 3-axis gyroscope, 3-axis accelerometer and 3-axis magnetometer. Pedestrian position estimation can be achieved based on IMU. Theoretically, by integrating the measured accelerations twice, the moving distance can be calculated. The orientation information can be obtained based on the measurements from the magnetometer and gyroscope. However, the calculated distance drifts away in a short period of time due to the biases in the acceleration measurements. Because of the drift errors, the IMU localization accuracy is not so promising. To improve this performance, Pedestrian Dead-Reckoning (PDR) is developed and used as an infrastructure-free methodology for pedestrian localization based on IMU.

The IMU is mounted on the foot of a human. The basic steps of the PDR algorithm are step counting; orientation information computation; step length calculation; finally, position estimation. The step counting can be realized by detecting the stance and swing phase of the foot based on the variance of the accelerations or the angular velocities, etc. [ZLWW17].

The calculation of the orientation (e.g. Euler angles or quaternion) is achieved based on the measurements from magnetometers and gyroscopes. By integrating the acceleration, the step length can be obtained. During the stance phase of each step, the foot velocities should be zero after the integration of the acceleration. However, since biases exist in the accelerations, the foot velocities are not zero. Theoretically, the calculated velocities in the stance phase are equal to the integration of the biases. Thus, the biases can be calculated using these velocities. Based on this principle, the drift error can be reduced. The position estimation can be achieved with the calculated step numbers, step lengths and orientation information.

1.1.3 Simultaneous Localization and Mapping (SLAM)

In general, simultaneous localization and mapping (SLAM) is a technique for building a map of an unknown environment and estimating positions. This technology has been used in many areas, such as in robots, autonomous cars, unmanned aerial vehicles, and augmented reality (AR). Lidar and cameras are the most widely used sensors for SLAM.

Compared to camera based SLAM, localization based on light detection and ranging (Lidar) is more accurate. The distance between the target and the Lidar can be measured with the help of a laser.

On the other hand, using a camera is cheaper and can provide more visual information compared to Lidar. There are three different kinds of cameras for SLAM application: monocular, stereo and RGB-D cameras. A monocular camera contains only a single camera, while a stereo camera has two cameras. Besides an RGB image, the distance between the camera and the object is also provided by the RGB-D camera. Camera based SLAM can be realized with four steps: visual odometry, back end, loop closing and mapping.

1) Visual odometry (VO): The translation and rotation between adjacent frames are determined and used as initialization values for the back end. There are three different methods to calculate the camera motion: the feature based method [MMT15], optical

flow [WCD+16] and direct method [FPS14].

2) Back end: The camera motions are optimized with frames taken at different times based on the extended Kalman filter, bundle adjustment, pose graph, and so forth [SP12].

3) Loop closing: The main task of loop closing is to identify whether the camera has returned to the previous area.

4) Mapping: Based on the estimated camera motions, the map can be built.

1.1.4 Visible Light or Ultrasound Based Indoor Localization

Position estimation can also be realized based on visible light. The most widely used element in the visible light system is the light-emitting diode (LED). The TOA, TDOA, RSS and AOA position estimation algorithms can also be used in an LED localization system. A method is proposed in [MSM08] to determine the position and the direction of the receiver based on visible LED lights and image sensors.

Sound waves with frequencies higher than 20 kHz (upper limit frequency of human hearing) are defined as ultrasound. Like wireless systems, the TOA and TDOA algorithms are also applied in ultrasound systems for localization purposes.

1.2 Comparison of Localization Systems

Each of the presented localization systems has its own advantages and disadvantages. Several factors can be considered to select the most suitable system for different indoor applications, such as cost, position estimation accuracy, update rate, system capacity, coverage area, system complexity and power consumption. The position estimation frequency of the system per second is defined as the update rate. The costs for indoor localization systems can differ widely. Normally, the more accurate the system is, the more expensive it is. The system capacity determines how many devices can be used at the same time. For the RSVP project application, the following factors are the most important in selecting the localization system.

1) Accuracy: The position estimation accuracy is the most important factor to guarantee human safety. The system should be able to continuously provide the precise location of humans and robots so that the distances between them can be monitored with a relatively high frequency. Furthermore, the more accurate the system is, the smaller the SSD and LSD zones are and the more space can be saved for further applications.

2) Update rate: When humans move fast, if the update rate is too low, the human could already be in the danger zone, and the real-time distance may not be updated yet or the robot may move too fast and not be stopped in time. This could be very dangerous for the workers.

3) System capacity: In the production line, the number of workers and robots could be more than 100. Thus, the system must be able to provide enough devices, that can be used for localization at the same time.

4) System complexity: To maintain the flexibility of the system, its complexity should be low. The easy installation and easy adding or removing localization devices after the installation are required.

5) Coverage: The system should be able to cover the whole working area. In the RSVP project, the coverage should be at least 25 m.

6) Cost: The cost of the system needs to be kept in an acceptable range.

The comparison of the localization accuracy, coverage area, complexity and cost of different systems can be found in Table 1.1. As shown in the table, compared to Wi-Fi, Bluetooth, ZigBee and RFID, UWB has better accuracy and its coverage meets the requirements. The SLAM based solutions are even more accurate than UWB, but the cost of hundreds of Lidar SLAM devices is much higher. Due to the confidential environments in the production line, cameras are not a good option. Furthermore, once the direct signal propagation path is blocked, many localization systems become inaccurate. However, UWB can still provide accurate measurements if the blockage only comes from, for instance, thin tables, counters or glass. The update rate of UWB can exceed 90 Hz, which is more than enough for the RSVP project. With the developed UWB system in the project, 100 localization MSs with an update rate of 20 Hz can be used at the same time. The cost of the system is acceptable. Thus, the UWB localization system is the best option for the RSVP project application.

1.3 General Information on UWB

Despite the advantages of UWB, some challenging problems remain for UWB based indoor localization, such as the inaccurate position estimation caused by NLOS errors. First, this section presents a short introduction to the UWB system. Due to the fine time resolution of the UWB signals, the time based TDOA and TOA stands in the focus in this thesis, since they offer very good accuracy. The basic principle of time based UWB localization can be explained as follows. The MS sends the UWB pulse to the BSs. The BSs receive the signals from direct paths, reflected paths, etc. and then provide the channel impulse response(CIR). Given the CIR, the arrival time can be obtained. With the help of the sending time from the MS, the signal propagation

Table 1.1: Comparison of different localization systems [Liu14], [XZYN16], [Mau12], [MPS14], [LDBL07] , [ZGL17]

	Accuracy	Coverage (m)	Complexity	Cost
Wi-Fi	m	20-50	low	low
UWB	cm-m	1-50	low	low/medium
Infrared (IR)	cm-m	1-5	low	medium
Bluetooth	m	10	low	low
ZigBee	m	30-60	low	low/medium
RFID	dm-m	1-50	low	low
IMU	1%	10-100	low	low
Vison	0.1mm-dm	1-10	high	high
Ultrasound	cm	2-10	low	low

time from the MS to the BSs can be calculated. Based on these time measurements, the range (for TOA) or the range difference (for TDOA) can be obtained and used for position estimation. However, these data suffers from noise errors and NLOS errors. Inaccurate localization is mainly caused by NLOS errors. Different methods have been developed for NLOS mitigation, such as the Kalman filter (KF) and particle filter (PF), which are combined with CIR based NLOS identification, IMU sensor based NLOS mitigation and so forth. In summary, UWB based localization can be divided into the following parts, as show in Figure 1.5:

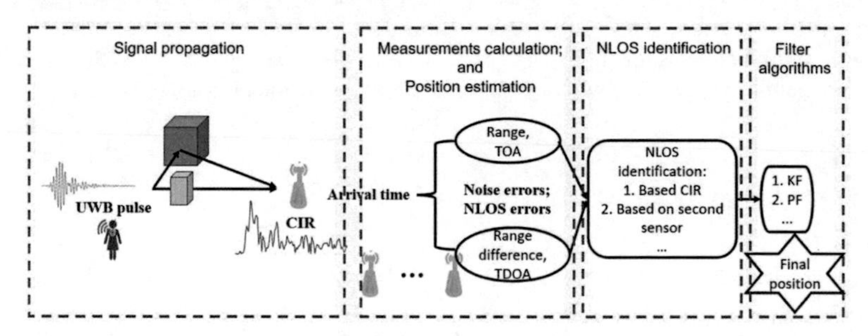

Figure 1.5: UWB based localization process

1. Signal propagation: The MS sends the signals and the BSs obtain the signals. The final results in this phase are the CIRs provided by the BSs.

2. Measurements calculation: Based on the CIRs, the signal arrival time can be ob-

tained. With the help of the arrival time and signal sending time, the range (for TOA) or the range difference (for TDOA) can be calculated.

3. Position estimation: Based on TOA or TDOA, the final position can be computed.

Additional methods to improve the localization accuracy are the following:

4. NLOS identification: NLOS identification can be realized using CIRs or a second sensor source, such as IMU, etc.

5. Filter algorithms: KF and PF.

UWB signal propagation and channel estimation have been discussed in many papers [LDM02], [CLLW16], [CSW02], [WRSB97], [WS02]. However, these propagation models are mostly for communications. Few papers include a detailed discussion about the relationships between CIRs and accurate/inaccurate measurements. Measurement calculation with one-way range or two-way range is described in [Utt15], [Decb]. Based on the different types of measurements (range, range difference, etc.), different methods have been developed for position estimation, such as TOA, TDOA, AOA and RSS, as discussed in Section 1.1.1. The main factor that causes inaccurate localization is the NLOS error. NLOS identification is the most powerful approach to reduce this error. The overview of NLOS identification and mitigation is provided in chapter 4. Although UWB localization systems have been discussed in many papers, there are still some unanswered or not fully answered questions. From the influence of the signal propagation path for the CIR to the final error mitigation algorithms, this thesis aims to provide a detailed overview of UWB based localization. As shown in Figure 1.6, the answers to the following questions in different phases for the UWB based localization process are given in this thesis:

During signal propagation:

1. How do the channel paths affect the CIRs?

2. Do all blockages cause inaccurate measurement?

During measurement calculation:

1. What does the LOS/NLOS noise distribution look like?

2. What is the relationship between the CIRs and NLOS/LOS measurements?

During position estimation:

1. What is the difference between TDOA and TOA?

2. Under which conditions is TDOA more suitable than TOA?

During NLOS identification:

1. What are the different NLOS identification methods?

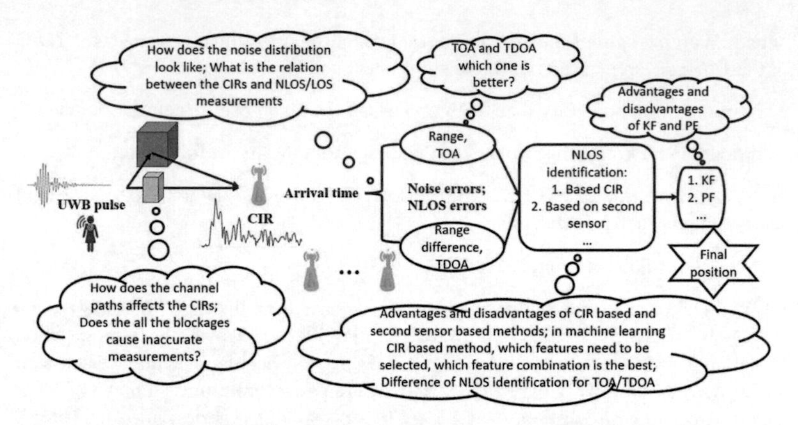

Figure 1.6: Overview for the UWB based localization process

2. What are the advantages and disadvantages of CIR based and second sensor based methods?

3. In machine learning CIR based methods, which features need to be selected, and which feature combination is the best?

4. What is the difference in the NLOS identification for TOA and TDOA?

During filter algorithms:

1. What are the advantages and disadvantages of KF and PF for UWB based indoor localization applications?

Based on the answers to these questions, this thesis shows that NLOS identification is the most effective method to improve UWB localization accuracy. Thus, NLOS identification stays in the focus in this thesis.

1.4 NLOS Identification for UWB

As described above, UWB based localization is not new. The range based TOA, range difference based TODA, angle based AOA and received signal strength based RSS position estimation algorithms for UWB have been discussed in many papers. Different filters, such as the Kalman- and particle filter, can be used to reduce the system random error. Under line of sight condition, where no blockage exists between the MS and BS, accurate UWB based localization can be achieved. The challenge arises if the signal propagation path is blocked, which is defined as NLOS. A non-negligible bias is added to the measurement under NLOS conditions, which leads to inaccurate position estimation if the NLOS measurement is used for localization calculation. The accurate NLOS identification is the key factor to guarantee UWB based accurate position estimation. Most of the current approaches can be summarized as follows.

Without a detailed investigation of the LOS/NLOS error, many approaches assume that the errors model can be treated as a Gaussian distribution, and they use the Kalman filter to identify and mitigate the NLOS errors. However, NLOS errors differ with different blockage materials. These errors can be any random values. Thus, a Gaussian distribution is not a good description of the NLOS error model.

Some papers assume that the variance of the NLOS measurements is theoretically larger than that of the LOS measurements. By using a suitable threshold to compare with the variances, NLOS detection can be achieved. However, if the blockage is the same and the distance between the BS and MS does not change dramatically, the variance of the NLOS errors can be smaller than the variance of the LOS errors. Thus, NLOS detection accuracy is not very promising. Besides, the proper threshold is highly difficult to be determined, and the latency can not be avoided due to the collection of the ranges for the variance calculation.

Other methods utilize additional information for the NLOS detection, such as RSS or a map of the floor. However, RSS is highly dependent on the indoor environments. With humans coming in or going out, the RSS could change dramatically. Besides, the threshold used to compare with the RSS is highly difficult to be determined. With an improper threshold, the accuracy can be very poor. The main disadvantage of the map based method is that the maps are not always available. Except for the fixed furniture, there are many of moving objects or humans. The NLOS detection accuracy based on these methods is not sufficient.

One of the most effective methods of NLOS identification is based on CIR. The current CIR based NLOS detection works only for TOA. Furthermore, the reasons for selecting the useful features have not been explained, and the optimization of the feature combination and the parameters in machine learning algorithms has not been discussed.

In summary, the current NLOS identification and mitigation methods have the following drawbacks:

1) The error model of the UWB measurements has not been discussed in detail. Thus, improper models are assumed and used for identification.

2) The relationship between the CIRs and the accurate/inaccurate range measurements has not been fully described. Not all blockages lead to inaccurate measurements. A detailed description of the conditions that cause the NLOS is still missing.

3) An overview and comparison of the current NLOS identification methods are miss-

ing.

4) Although TOA and TDOA are discussed in many papers, they are barely compared based on the localization environment perspective. The range measurements are used for localization in the TOA method. Under the condition that the NLOS condition can be properly identified, this method works very well in the environments where NLOS does not occur very frequently, such as offices. However, in harsh industrial environments where NLOS happens frequently, even with the help of the NLOS identification approach, this method might not provide accurate position estimation since it can happen that not enough accurate ranges can be obtained. In these environments, TDOA which used the range difference for position estimation can be more accurate than TOA, since the biases of two different NLOS ranges can be compensated and an accurate range difference can be obtained even under NLOS condition. If the accurate range difference can be selected, accurate localization can be guaranteed. However, most current approaches only discuss NLOS detection for the selection of the accurate ranges, and do not work well for the accurate range difference selection in TDOA.

5) The fusion of IMU and UWB is one of the most widely used methods to improve localization accuracy. Most current fusion approaches are based on the extended Kalman filter with the assumption that the errors are Gaussian distributed. The NLOS outliers are detected based on predicted Gaussian distribution error models. However, in reality, NLOS errors are not Gaussian distributed.

To further improve the NLOS identification and UWB based localization accuracy, in this thesis, the UWB measurement errors are investigated. LOS and NLOS errors are discussed separately. LOS errors are evaluated in office and industrial environments with different distances. For NLOS errors, different blockage materials and blockage conditions are considered. The LOS and NLOS error models can only be built based on the experimental investigation since the theoretical modeling can not be achieved in different test environments. Systematic experimental investigation on NLOS effects have been done in office and industrial environments. The relationship between CIRs and accurate/inaccurate range measurements is described in detail. Furthermore, the thesis presents an overview of the current NLOS identification methods. Four different NLOS identification and mitigation approaches are developed in this project. The first one works for the TOA approach in an office environment with a stand-alone UWB system, while the second one works in harsh industrial environments. The third and fourth one are UWB/IMU fusion approaches, and they work for TOA/TDOA methods.

1.5 Outline and Contributions

To further improve NLOS identification and localization accuracy, the following investigations are conducted.

First, the UWB system is systematically analyzed, and the possibility to build general error models based on collected UWB measurements is evaluated. It is found that although the noise distribution under LOS can be modeled as a Gaussian distribution, a general error distribution model under NLOS is difficult to build due to the unstable NLOS error. The error model is different depending on the environment. Specifically, for the Bosch Shanghai office, an approximate stable distribution model can be used to describe the error distribution based on our investigation. Although the localization performance can be improved with properly built error distribution models, NLOS measurements still have an influence on the position estimation accuracy in the field test.

Next, the thesis describes UWB signal propagation in detail. The relationship between the CIRs and the accurate/inaccurate range measurements is theoretically discussed in three different situations: ideal LOS path, small-scale fading: multipath, and NLOS path. The theoretical relationship is validated with real measured CIRs in Bosch Shanghai office environment. It is found that not all blockages lead to inaccurate measurements. The thesis explains when and why the blockages cause inaccurate measurements.

Next, the thesis presents an overview of NLOS identification for the TOA method. It shows that CIR based- and second sensor based NLOS detection are two of the best approaches from the perspective of the NLOS detection accuracy and engineering feasibility. For the current CIR based NLOS detection, a summary of the features and the optimization of the feature combination as well as the parameters in machine learning are missing. In this thesis, an overview of the possible useful features are given. Based on the difference in CIRs of the accurate and inaccurate range measurements, five different feature groups are created according to distance, CIR shape, time, multipath richness and power related features. In each group, several features can be extracted. With these features, NLOS identification is realized based on the SVM method. The optimal feature combination is theoretically determined and validated with real measurements. The parameters in SVM are optimized. The localization accuracy shows highly promising improvements based on the particle filter with the NLOS identification compared to the traditional methods in the Bosch Shanghai office environment.

The thesis then presents the difference between the TOA and TDOA. It shows that in harsh industrial environments, where NLOS conditions frequently occur, the localization accuracy improvement based on particle filter with the NLOS identification TOA are limited. A novel approach is proposed which combines TOA and TDOA method with accurate range and range difference selection. The position estimation accuracy is improved with this approach compared to the other approaches during a field test in the Bosch Changsha plant.

Different to the current UWB/IMU fusion approaches, which detect inaccurate UWB measurements based on the assumption that the errors are Gaussian distributed, in this paper, the triangle inequality theorem is used to select the accurate ranges for TOA or the accurate range differences for TDOA based on the IMU measurements. The Gaussian distributed error models are not needed for the proposed approaches. The position estimation accuracy is improved with the proposed approach compared to the traditional methods in the Bosch Shanghai office.

1.5.1 Contributions

In summary, the main contributions of this thesis are follows:

1) This thesis investigates LOS/NLOS errors in different environments under different situations. Based on the investigation, an approximate stable distribution model is built to describe the error distribution in the Bosch Shanghai office.

2) The relationship between the CIRs and the accurate/inaccurate range measurements is theoretically discussed and validated with the real measured CIRs in the Bosch Shanghai office environment. The thesis explains when and why the blockages cause inaccurate measurements.

3) The extracted features from CIR for NLOS detection are divided into five different groups. The optimal feature combination is theoretically determined and validated with real measurements. Furthermore, the parameters in SVM are optimized.

4) The difference between TOA and TDOA is discussed from the localization environment perspective. Based on this discussion, a novel approach is proposed, which combines TOA and TDOA methods with accurate range and range difference selection.

5) The thesis proposes two UWB/IMU fusion approaches that utilize the triangle inequality theorem to select accurate ranges for TOA or accurate range differences for TDOA based on IMU measurements. Gaussian distributed error models are not needed for the proposed approaches.

1.5.2 Outline

This thesis is organized as follows.

Chapter 1 briefly introduces the "Real-time Safety Virtual Positioning" (RSVP) project and provides an overview of the existing indoor localization systems. After the comparison of these position estimation systems, the UWB system is determined to be suitable for the project. A short overview of UWB is provided. Finally, the outline and main contributions of this thesis are described. This chapter is partly based on the following papers:

W. Wang, Z. Zeng, W. Ding, H. Yu and H. Rose, "Concept and Validation of a Large-scale Human-machine Safety System Based on Real-time UWB Indoor Localization*," 2019 IEEE/RSJ International Conference on Intelligent Robots and Systems (IROS), Macau, China, 2019

Z. Zeng, L. Wang and S. Liu, "An introduction for the indoor localization systems and the position estimation algorithms," 2019 Third World Conference on Smart Trends in Systems Security and Sustainablity (WorldS4), London, United Kingdom, 2019

Chapter 2 describes the UWB system in detail. It provides an overview of the factors that influence UWB localization accuracy. These factors are the antenna, installation of the BSs, time synchronization, localization algorithms, NLOS/LOS identification, and filter algorithms. The signal propagation path is introduced. The measurement error is investigated. The relationship between CIRs and accurate/inaccurate measurements is theoretically explained for three different cases: ideal LOS path; small-scale fading: multipath; NLOS path. Furthermore, the chapter investigates the UWB range measurements error under clear LOS/multipath LOS, ignorable NLOS blockage and non-ignorable NLOS blockage.

Chapter 3 focuses on the localization filter problem. The Kalman filter principle is presented. It shows that Kalman filter can provide estimated state vectors with minimized variances for linear problems with Gaussian noise. Extended Kalman filter at the same time can be used for non-linear problems with Gaussian noise. However, based on the experiment results, it might suffer from a divergent problem with the frequently changing measurement noise. IEKF is more stable compared to EKF. The particle filter can be used to solve non-linear problems with non-Gaussian noise. PF is more stable compared to IEKF based on the experiment results. However, PF has the highest computation load.

Chapter 4 provides an overview of the NLOS identification methods: these methods are based on range variance estimation, combination of RSS, map, CIR, CIR state change and IMU. The chapter compares these methods' identification accuracy, engineering feasibility, and so forth. After the comparison, it can be determined that CIR based- and IMU based NLOS identification are two of the best approaches.

Chapter 5 focuses on the UWB localization in office environments. First, the localization with three BSs is discussed. The error distribution is found to be stable distribution. Since the system is non-linear and the error distribution is not Gaussian, the PF is used for position estimation. Although it can be observed that the localization accuracy is improved with a properly defined error distribution, the NLOS error is still severe. A better solution to further improve the accuracy is to add redundant BSs and use only the selected accurate ranges based on NLOS identification for further calculation. The SVM algorithm is used for NLOS detection. Five different feature groups are divided based on the distance, CIR shape, time, multipath richness and power related features. The reasons why these features can be used for classification are explained.

The feature combination, the used CIR length and the parameters in SVM are optimized to improve the identification accuracy. It can be observed in the field test that, the localization accuracy with NLOS identification is dramatically improved.

Chapter 6 presents UWB localization in harsh industrial environments. NLOS conditions occur more frequently in harsh industrial environment than in office environments. In the Bosch Changsha plant, it often happens that less than two ranges are measured under LOS. The position estimation with the NLOS identification based TOA approach does not have very good accuracy. Hence, an accurate ranges and range differences identification based TOA/TDOA combination approach is proposed to improve position estimation accuracy. Two SVM models are trained. The first one is used to select the accurate ranges, which is the same as the one presented in Chapter 5. If at least three ranges are detected as accurate, then the localization can be realized with these ranges. Otherwise, the range differences need to be calculated with the inaccurate ranges. The second SVM model is used to select the accurate range differences. These accurate ranges and range differences are used for position estimation. The particle filter is used to realize the localization together with the accurate ranges and range differences identification based TOA/TDOA combination approach. The position estimation with the proposed TOA/TDOA combination approach shows better accuracy than the NLOS identification based TOA approach and the standard TOA approach in the Bosch Changsha plant.

Chapter 7 focuses on UWB/IMU fusion localization. The IMU measurements are used to identify and mitigate UWB NLOS errors. The fusion system can be used for both TOA and TDOA based on UWB localization. With the help of IMU measurements, the accurate ranges for TOA or the accurate range differences for TDOA can be determined based on the triangle inequality theorem. The localization performance with the UWB/IMU fusion system is evaluated in the Bosch Shanghai office. The localization accuracy is improved with the proposed methods compared to the methods without the fusion with IMU. This chapter is based on the following papers:

Z. Zeng, S. Liu, and L. Wang. Uwb/imu integration approach with nlos identification and mitigation. In 2018 52nd Annual Conference on Information Sciences and Systems (CISS), pages 1-6, March 2018.

Z. Zeng, S. Liu, and L.Wang. A novel nlos mitigation approach for tdoa based on imu measurements. In 2018 IEEE Wireless Communications and Networking Conference (WCNC), pages 1-6, April 2018.

Chapter 8 summarizes this thesis and provides the conclusions.

1.6 Publications

The following papers are published during the PhD studies:

Z. Zeng, S. Liu, W. Wang and L. Wang, "Infrastructure-free indoor pedestrian tracking based on foot mounted UWB/IMU sensor fusion," 2017 11th International Conference on Signal Processing and Communication Systems (ICSPCS), Surfers Paradise, QLD, 2017, pp. 1-7. doi: 10.1109/ICSPCS.2017.8270492

Z. Zeng, S. Liu and L. Wang, "NLOS Identification for UWB Based on Channel Impulse Response," 2018 12th International Conference on Signal Processing and Communication Systems (ICSPCS), Cairns, Australia, 2018, pp. 1-6. doi: 10.1109/ICSPCS.2018.8631718

Z. Zeng, S. Liu and L. Wang, "UWB/IMU integration approach with NLOS identification and mitigation," 2018 52nd Annual Conference on Information Sciences and Systems (CISS), Princeton, NJ, 2018, pp. 1-6. doi: 10.1109/CISS.2018.8362197

Z. Zeng, S. Liu and L. Wang, "NLOS Detection and Mitigation for UWB/IMU Fusion System Based on EKF and CIR," 2018 IEEE 18th International Conference on Communication Technology (ICCT), Chongqing, 2018, pp. 376-381.

Z. Zeng, S. Liu and L. Wang, "A novel NLOS mitigation approach for TDOA based on IMU measurements," 2018 IEEE Wireless Communications and Networking Conference (WCNC), Barcelona, 2018, pp. 1-6. doi: 10.1109/WCNC.2018.8377041

Z. Zeng, S. Liu and L. Wang, "UWB NLOS identification with feature combination selection based on genetic algorithm," 2019 IEEE International Conference on Consumer Electronics (ICCE), Las Vegas, NV, USA, 2019, pp. 1-5.

Z. Zeng, R. Bai, L. Wang and S. Liu, "NLOS identification and mitigation based on CIR with particle filter," 2019 IEEE Wireless Communications and Networking Conference (WCNC), Marrakesh, Morocco, 2019, pp. 1-6, doi: 10. 1109 / WCNC. 2019. 8886002.

W. Wang, Z. Zeng, W. Ding, H. Yu and H. Rose, "Concept and Validation of a Large-scale Human-machine Safety System Based on Real-time UWB Indoor Localization*," 2019 IEEE/RSJ International Conference on Intelligent Robots and Systems (IROS), Macau, China, 2019, pp. 201-207, doi: 10.1109/IROS40897.2019.8968572.

Z. Zeng, L. Wang and S. Liu, "An introduction for the indoor localization systems and the position estimation algorithms," 2019 Third World Conference on Smart Trends in Systems Security and Sustainablity (WorldS4), London, United Kingdom, 2019, pp. 64-69, doi: 10.1109/WorldS4.2019.8904011.

Z. Zeng, W. Yang, W. Wang, L. Wang and S. Liu, "Detection of the LOS/NLOS state change based on the CIR features," 2019 Third World Conference on Smart Trends in Systems Security and Sustainablity (WorldS4), London, United Kingdom, 2019, pp. 110-114, doi: 10.1109/WorldS4.2019.8904000.

2 Analysis of UWB Based Localization System

UWB contains three main components: the mobile station (MS), base station (BS), and software to realize the localization/filter algorithms, as shown in Fig. 2.1.

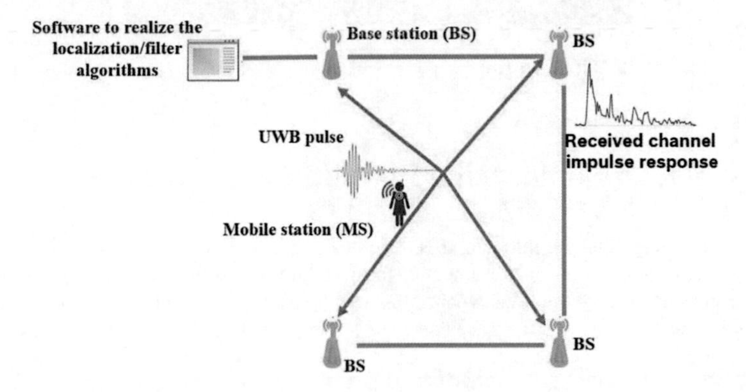

Figure 2.1: UWB based localization system architecture

MS: The MS is attached to the human or robots. It sends the UWB pulse to the BSs.

BS: Due to the signal reflection and diffraction etc., the BS receives not only the signal sent by MS, but also the reflected signal. The sum of these signals is the channel impulse response (CIR). Based on the CIR, the signal propagation time can be obtained. A detail discussed about the UWB signal propagation and CIR can be found in section 2.3. With the help of the signal propagation time, the arrival angle or the received strength can be obtained. Based on the type of measurements, different localization algorithms can be used (e.g. TOA, TDOA). For the accurate position estimation, the fixed installation of the BSs is needed, and the positions of the BSs need to be determined at the very beginning.

Software: The software is used to process the measurement data. The localization algorithms and the filter algorithms are realized in the software. The final position is calculated and provided by the software.

The hardware design of the used UWB localization system is shown in Fig. 2.2.

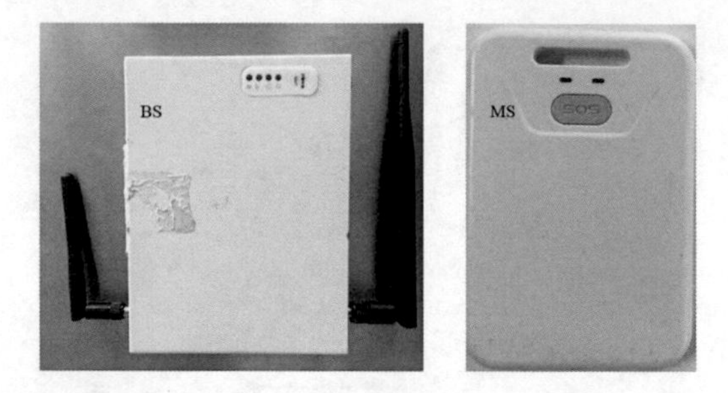

Figure 2.2: The BS (left side of the figure) and the MS (right side of the figure)

2.1 Factors Influencing Localization Accuracy

There are many factors, that can affect the position estimation accuracy, such as the type or the orientation of the antenna, the installation position of the BSs, the localization algorithm, the time synchronization, NSLOS/LOS identification, and filter algorithms. An overview for these factors is presented in Fig. 2.3 and discussed in this chapter.

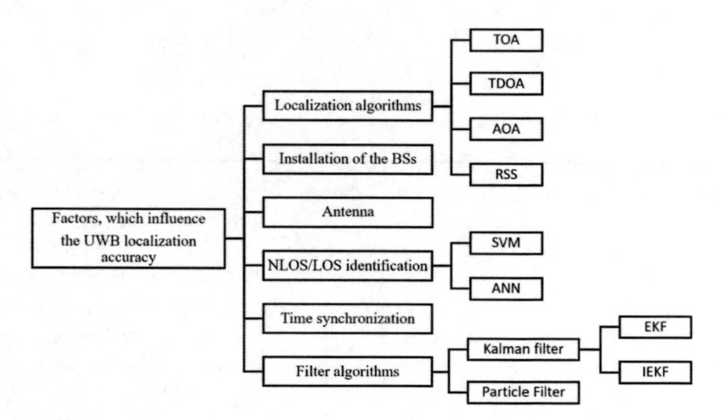

Figure 2.3: Factors, that influence UWB localization accuracy

Antenna

An omni-directional antenna is used in the UWB system. Although the antenna is omni-directional, its orientation in the BS still has an influence on the range measurements. Three different situations are investigated here: the angle of the BS's antenna changes from 0 to 45 degrees, and then to 90 degrees, while the rest of the conditions, such as the position of the MS, the distance (5 m) between the BS and the MS, the height of the MS and BS and the environments, are kept the same, as shown in Fig. 2.4. 300 tests for each angle of the antenna are done under the same condition. One example of the error distributions for different antenna angles are presented in Fig. 2.5. In total 10,000 errors are collected in each test. As shown in this figure, the mean error with the 0 degree angle is the smallest. The variances of the 90 degree and 0 degree are not much different. Based on our observations for the errors with different angles in these tests, the orientation of antenna still influences the measurements accuracy, even though the antenna is omni-direcitonal. In order to find the best angle for position estimation, the position of the BS is fixed and the antenna angle is either perpendicular or at an angle of 45, 90 degrees to the ground. The MS is placed in 100 different positions on the ground in the office. There is no blockage between the MS and BS. The reference distance between them is measured by the Bosch GLM 100 C Professional laser measure. Based on these experiments, if the antenna of the BS is perpendicular to the ground, the average error and the standard deviation are the smallest. Thus, during the BS installation, the antennas of the BSs are all perpendicular to the ground. The errors caused by the antenna are considered as LOS noise errors.

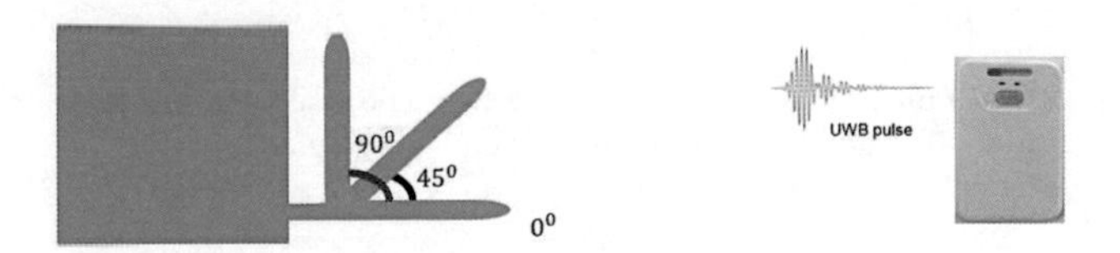

Figure 2.4: Orientation of the antenna

Installation Position of the BSs

To realize accurate position estimation, the BSs need to be fixed in a known position. If the BSs are installed in the corner, their antennas have to be kept at a certain distance to the wall to avoid the influence of reflected signals from the wall.

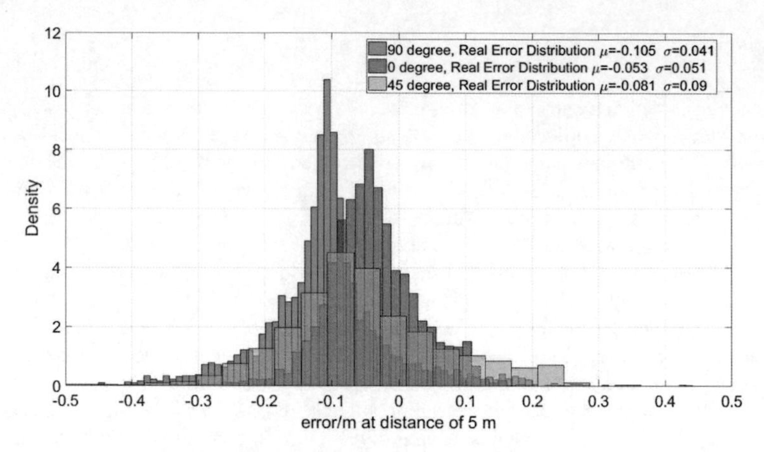

Figure 2.5: Error distribution for different antenna angles

Time Synchronization

For time based TDOA and TOA, time synchronization is crucial. For two way ranging TOA, the synchronization between BSs and the MS are not required [YWO13], [CCGZ06], [HIM17]. However, for TDOA, the synchronization between BSs is necessary. Thus, during the testing, it is found that TOA is generally more accurate than TDOA, since the synchronization problem between BSs for TDOA causes additional errors. However, in harsh industrial environments, the TDOA can be more accurate than the TOA. The main reason is that, due to the compensation of the errors in the arrival times, the obtained range difference measurements under NLOS conditions can still be accurate. This situation is discussed in detail in Chapter 6.

Localization Algorithm

As described in the introduction, the TOA, TDOA, AOA and RSS localization algorithms are used for position estimation in the UWB system. Thanks to the fine time resolution of the UWB signals, time based TDOA and TOA are the focus of this thesis, since they offer very good accuracy. The geometric explanation of TOA and TDOA was presented in the introduction. The intersection point of the circles in TOA and the intersection point of the hyperbolas in TDOA is the position of the MS. The Taylor series algorithms can be used to realize the TOA and TDOA approach.

a) Taylor series algorithms for TOA: In TOA, range measurements are used for position estimation. The range between the MS and the ith BS can be represented with the following equation [GZT08b]:

$$D_{i,t} = \sqrt{(x_i - x_t)^2 + (y_i - y_t)^2 + (z_i - z_t)^2} \tag{2.1}$$

where the position of the ith BS is represented by $X_i = (x_i, y_i, z_i)$, and $X_s = (x_t, y_t, z_t)$ is the real position of the MS. z_i and z_t are constant, since the height of the MS and BS is fixed.

Based on Taylor series algorithms, the range function can be expanded into Taylor series for linearization purposes:

$$D_{i,t} \approx D_{i,t-1} + \frac{\partial D_{i,t-1}}{\partial x}\delta_x + \frac{\partial D_{i,t-1}}{\partial y}\delta_y + \frac{\partial D_{i,t-1}}{\partial z}\delta_z \tag{2.2}$$

where

$$\frac{\partial D_{i,t-1}}{\partial x} = \frac{x_i - x_{t-1}}{\sqrt{(x_i - x_{t-1})^2 + (y_i - y_{t-1})^2 + (z_i - z_t)^2}} \tag{2.3}$$

$$\frac{\partial D_{i,t-1}}{\partial y} = \frac{y_i - y_{t-1}}{\sqrt{(x_i - x_{t-1})^2 + (y_i - y_{t-1})^2 + (z_i - z_t)^2}} \tag{2.4}$$

$$\delta_x = x_t - x_{t-1} \tag{2.5}$$

$$\delta_y = y_t - y_{t-1} \tag{2.6}$$

$$\delta_z = z_t - z_{t-1} = 0 \;\; (z_t \;\; is \;\; constant) \tag{2.7}$$

After the linearization of all the measured ranges, (2.2) can be rewritten in matrix form:

$$A\delta = D \tag{2.8}$$

$$A = \begin{pmatrix} \frac{\partial D_{1,t-1}}{\partial x} & \frac{\partial D_{1,t-1}}{\partial y} \\ \frac{\partial D_{2,t-1}}{\partial x} & \frac{\partial D_{2,t-1}}{\partial y} \\ \vdots & \vdots \\ \frac{\partial D_{n,t-1}}{\partial x} & \frac{\partial D_{n,t-1}}{\partial y} \end{pmatrix} \tag{2.9}$$

$$\delta = [\delta_x, \; \delta_y]^T \tag{2.10}$$

$$D = [D_{1,t} - D_{1,t-1}, D_{2,t} - D_{2,t-1}, \ldots, D_{n,t} - D_{n,t-1}]^T \tag{2.11}$$

where n is the number of used BSs.

Based on the weighted least squares method, δ can be calculated with the following equation:

$$\delta = (A^T W A)^{-1} A^T W D \tag{2.12}$$

$$W = \begin{pmatrix} w_1 & 0 & \ldots & 0 \\ 0 & w_2 & \ldots & 0 \\ \vdots & \vdots & \ldots & 0 \\ 0 & 0 & \ldots & w_n \end{pmatrix} \tag{2.13}$$

where w_i is the weight and can be calculated based on the signal noise of the ith BS. After the calculation of δ, the current position can be updated with the help of the previously calculated position.

b) Taylor series algorithms for TDOA: For TDOA, range difference measurements are used for localization. The range difference for the ith BS and the jth BS can be represented with the following equation:

$$\begin{aligned} D_{ij,t} = &\sqrt{(x_i - x_t)^2 + (y_i - y_t)^2 + (z_i - z_t)^2} \\ &-\sqrt{(x_j - x_t)^2 + (y_j - y_t)^2 + (z_j - z_t)^2} \end{aligned} \tag{2.14}$$

Again, the range difference function needs to be expanded into Taylor series with (2.2) for linearization purposes.

$$\begin{aligned} \frac{\partial D_{i,t-1}}{\partial x} = &\frac{x_i - x_{t-1}}{\sqrt{(x_i - x_{t-1})^2 + (y_i - y_{t-1})^2 + (z_i - z_t)^2}} \\ &-\frac{x_j - x_{t-1}}{\sqrt{(x_j - x_{t-1})^2 + (y_j - y_{t-1})^2 + (z_j - z_t)^2}} \end{aligned} \tag{2.15}$$

$$\begin{aligned} \frac{\partial D_{i,t-1}}{\partial y} = &\frac{y_i - y_{t-1}}{\sqrt{(x_i - x_{t-1})^2 + (y_i - y_{t-1})^2 + (z_i - z_t)^2}} \\ &-\frac{y_j - y_{t-1}}{\sqrt{(x_j - x_{t-1})^2 + (y_j - y_{t-1})^2 + (z_j - z_t)^2}} \end{aligned} \tag{2.16}$$

The rest of the calculation is the same as the Taylor series algorithms for TOA.

NLOS/LOS Identification

Although UWB based localization accuracy is highly promising under LOS, the estimated positions can not be trusted if the measurements are obtained under NLOS. A LOS path is a straight line signal propagation path that connects the MS and BS without any obstruction. If the direct signal propagation path is obstructed, the signal reaches the receivers with a time delay through the direct path or by reflected, diffracted or scattered paths; this is defined as NLOS. With NLOS signal propagations, accurate position estimation can still be achieved under three conditions: firstly, redundant BSs are available; secondly, the accurate measurements can be identified; and finally, only accurate measurements are used for further calculation.

In the TOA approach, the arrival time contains delays under NLOS conditions, causing inaccurate range measurements. At least three BSs are needed for position estimation. If redundant BSs (more than three BSs) are available, accurate localization can be achieved under NLOS by selecting the NLOS range measurements out based on NLOS identification algorithms.

For the TDOA approach, although the arrival time contains positive bias under NLOS, the biases for two different BSs can compensate with each other if they are equal. Under this condition, the range difference measurement can still be accurate with inaccurate arrival time. Four situations need to be considered: firstly, if both BSs are under LOS, the range difference is accurate; secondly, if only one of the BSs is obstructed, all the range differences calculated by two BSs containing this BS have a bias; thirdly, if it coincidentally happens that two arrival times contain the same delay, the range difference is accurate once again; and lastly, if two NLOS BSs contain different biases, the range difference is not accurate. Thus, it is more important to identify accurate range differences than to identify NLOS BSs.

A detailed description of NLOS identification in the TOA approach can be found in chapter 5, while the accurate range difference selection in the TDOA approach is covered in chapter 6.

Filter Algorithms

The Taylor series algorithm discussed above can be used for position estimation. However, Taylor series is a static model, which contains no internal history of the previous input, the previous estimated position or the errors. This means that the estimated position is solely dependent on the current measurements. The Taylor series does not describe the dynamic process of the localization. The previous estimated position, previous inputs and the error information are not used to optimize the current position results. However, the position estimation process is a dynamic process. The current position is dependent on the previous estimated position, previous inputs and the error information. To further improve accuracy, a dynamic process model of localization must be built. The initial position of the MS $(x_0\ y_0)$ at time t_0 can be calculated using the Taylor series algorithm. Based on the assumptions (e.g., the MS has constant velocity or acceleration) or measurements from a second sensor (e.g. outputs from IMU), the future position at time t_1 can be predicted as $(\hat{x}_1\ \hat{y}_1)$. Since the assumptions or the measurements contain uncertainty or system noise, the exact position can not be determined. If the uncertainty or system noise distribution model can be described, the distribution of the position can be obtained. At the same time, the UWB measurements also contain errors. Once the error distribution model of the UWB measurements can be described, the filters (e.g. Kalman filter) can be used to realize the optimal position estimation at time t_1 based on this distribution and the distribution of the position calculated with the assumptions as presented in Figure 2.6. The variance of the optimal position estimation should be smaller than that of the other distributions. After determining the optimal position at time t_1, the optimal position estimation at time t_2 can be achieved with the same steps. The optimal position estimation at time t_i can be

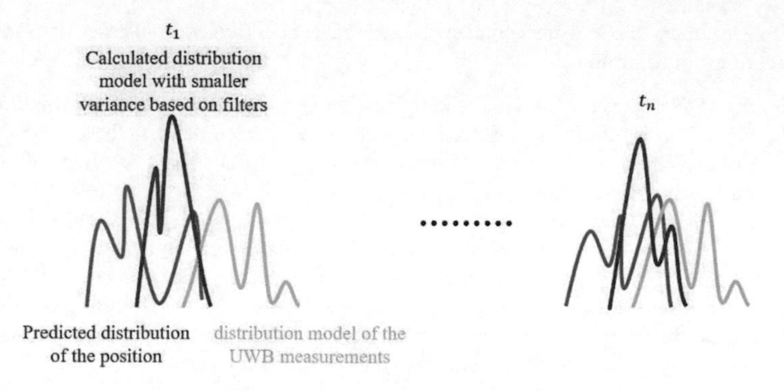

Figure 2.6: System dynamic process

obtained with the determined position estimation at time t_{i-1}.

Principally, in localization applications, filters are used to combine two distributions to obtain a new one with smaller variance and then determine the optimal position. In other words, filter algorithms are used to reduce the random system noise error and the NLOS error. These algorithms can dramatically improve accuracy with the combination of the NLOS identification/accurate measurements selection approaches. Many filter algorithms have been developed to improve accuracy, but this thesis focuses on the Kalman filter (KF) and the particle filter. Both filters describe a dynamic model of the system. The main advantage of the KF is its low calculation complexity. The disadvantage is that in order to obtain the optimal solution, the noise has to be Gaussian noise. On the other hand, the particle filter can deal with non-Gaussian noise, but the calculation complexity is higher compared to KF. The selection of different filters for different applications is important. Details about the localization dynamic process and the filters are discussed in chapter 3.

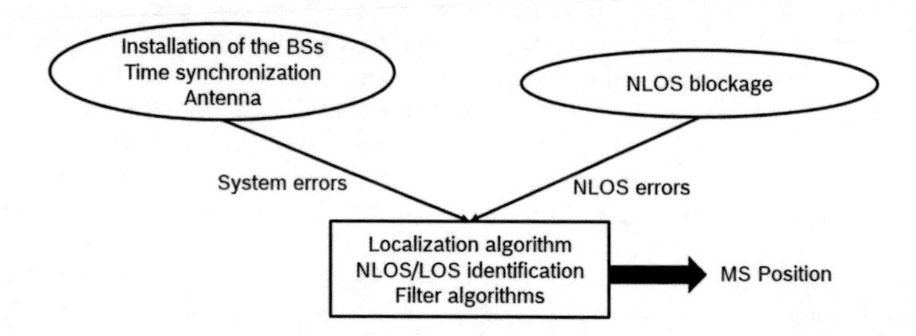

Figure 2.7: Relationship between the factors that influence accuracy

Relationship between these Influence Factors

The relationship between the factors discussed above is illustrated in Figure 2.7. System random errors are caused by the installation of the BSs, antenna and the time synchronization (for TDOA only), while the NLOS blockage is the main reason for the NLOS errors. The position estimation can be realized with TOA/TDOA. The NLOS/LOS identification and filter algorithms are used to reduce the system random errors and NLOS errors.

2.2 UWB LOS and NLOS Measurement Error

The error distribution of UWB measurements must be investigated since it is highly important for filter algorithms. The measurement error is defined as the difference between the measured range and the reference range, which are measured by the Bosch GLM 100 C Professional laser measure. The testing environment is the Bosch Shanghai office. Three different situations are considered: LOS; NLOS blockage with ignorable delay (e.g. glass, table, counters...); and NLOS blockage with non-ignorable delay.

Errors under Clear LOS

Although the errors under clear LOS (as shown in Figure 2.8) have roughly normal distribution, the mean and standard deviation of the distribution differ in different places, at different distances, or even with different MSs. Figure 2.8 shows the error distributions at the distances of 3, 5, and 6 m under the multipath LOS condition. The mean and standard deviation of the errors under the multipath LOS condition at different distances is presented in Table 2.1. The errors at different distances in different environments under clear LOS are illustrated in Figure 2.9. For each position, at least 10,000 ranges are collected. The observation of these errors shows that they change with the distance and the environment. Furthermore, under the same condition, with a different MS or BS, the error distribution also varies. The orientation of the antenna in the MS and BS influences the error as well. Although the error is not fixed, it is limited within 25 cm. The standard deviation also changes in an acceptable range. However, although the error distribution at a certain position is roughly a normal distribution, this distribution cannot be used to represent the error distribution for the whole office floor.

Table 2.1: Mean and standard deviation of the range errors at different positions

Real Distance (m)	Mean (m)	Error (m)	Deviation (m)
1.5	1.39	-0.11	0.092
2	1.88	-0.12	0.11
2.5	2.59	0.09	0.036
3	3.014	0.014	0.081
3.5	3.58	0.08	0.077
4	4.058	0.058	0.0633
4.5	4.48	-0.02	0.104
5	5.089	0.089	0.091
5.5	5.53	0.03	0.09
6	6.12	0.12	0.079
6.5	6.45	–0.05	0.082

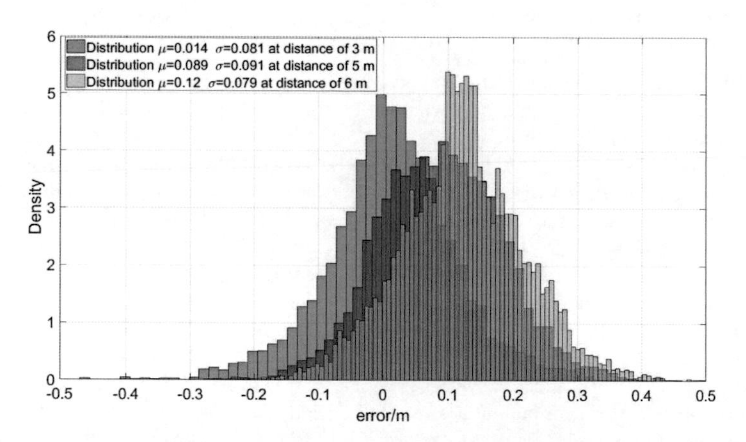

Figure 2.8: Error distribution at the distance of 3, 5 and 6 m

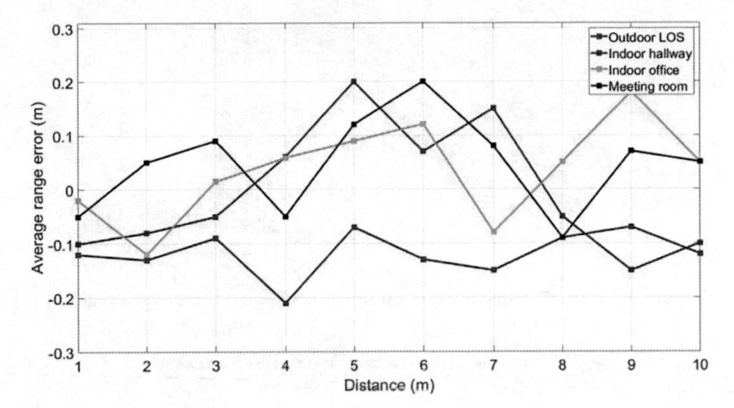

Figure 2.9: Error at different distances in different environments under clear LOS

Error under Ignorable NLOS Blockage

If the signal propagation is only blocked by a glass door, chairs or tables etc., the range measurement can still be accurate. During the data collection, the distance between the BS and MS is kept at 5.3 m, and different materials are used to block the signal propagation path. The measured distances under these blockages are presented in Figure 2.10. Even with a glass door, chair, table and counter blockage, the average error is still within 25 cm. The blockage error caused by these materials can be ignored. In this paper, this kind of ignorable NLOS blockage is defined as soft NLOS. During localization, the errors under LOS and under ignorable NLOS blockage are divided into the same group.

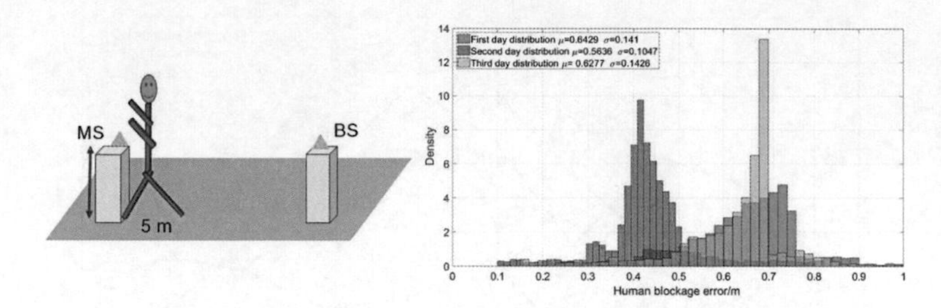

Figure 2.11: Error distribution for human blockage on three different days

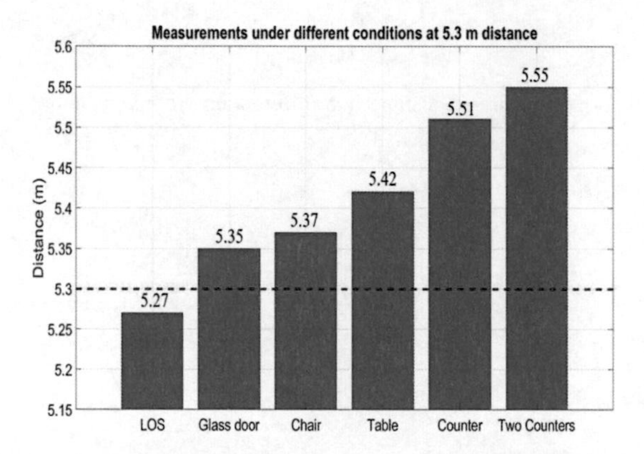

Figure 2.10: Measured distances under different ignorable blockages

Error under Non-ignorable NLOS Blockage

Inaccurate position estimation is mainly caused by non-ignorable NLOS blockage (such as human, water, metal, etc.). This kind of blockage is defined as hard NLOS (HNLOS). Two different situations are considered in this thesis to check whether a general model for the NLOS error can be built. In the first situation, the distance between the MS and BS is kept at 5m and the human stands before the MS in the same position on three different days. The error distribution in this case is presented in Figure 2.11. It can be observed in the figure that even with the same person creating the blockage, the error distributions are different.

In the second situation, the dynamic process is considered. First of all, the MS and BS are kept fixed at a distance of 3 m. Three cases are tested: firstly, thc human walks fast

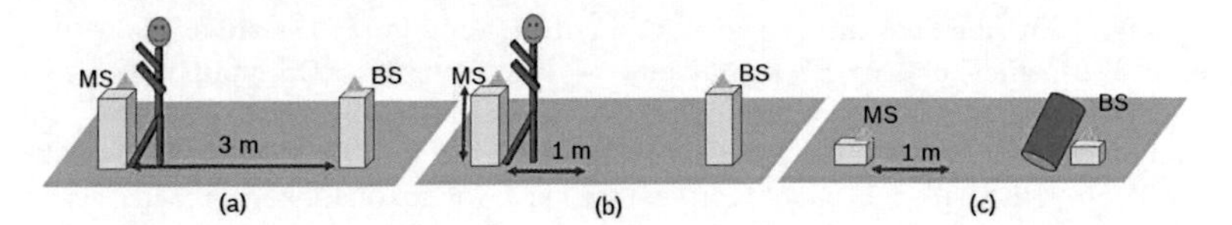

Figure 2.12: (a) Fast human walk; (b) slow human walk; (c) water blockage with moving MS

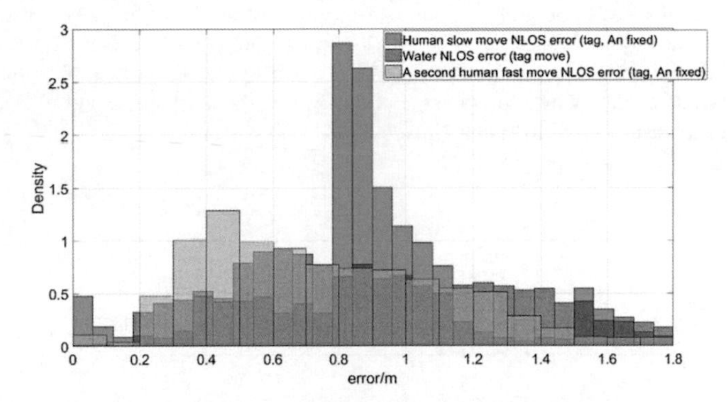

Figure 2.13: Error distribution for the dynamic process

between the MS and BS; secondly, the human walks slowly within 1 m distance at the MS's side; Thirdly, the water is located before the BS, and the MS moves within 1 m, as shown in Figure 2.12. The error distribution in these situations is shown in Figure 2.13. By comparing these figures, the error distributions changes under different conditions. Even under the same human blockage condition, the similarity of the distributions is low. Thus, a general NLOS error model with the developed UWB system cannot be built.

Summary

Although the error distributions under clear LOS and ignorable NLOS blockages at different positions differ, they can all be considered as roughly normal distributions. A Gaussian error distribution model can be built with these errors in different places. However, for different BSs, the error distribution models are different, as presented in Figure 2.14. Thus, for each BS, a separate error distribution model must be built.

However, for the NLOS error, a general model is highly difficult to build. If the heights of the BS and MS are fixed, at least three BSs are needed for position estimation.

Two different situations are considered. Firstly, if only three BSs are available, the error distribution for each BS can be built by combining the LOS and NLOS errors in different places, as shown in Figure 2.15. These distribution models can be built with stable distribution, as discussed in Chapter 5. Due to the rapid change of the NLOS error, these models fail to represent the real error distribution and the inaccurate measurements still have a huge influence on the final position estimation. Thus, the accuracy improvement with these models is limited in the field test. Secondly, more than three BSs are available, in other words, redundant BSs are used. If the NLOS measurements can be identified, they can be selected out and will not be used for position estimation. In this case, the NLOS error model is not necessary since it has no influence on the final position estimation. To realize the NLOS identification, it is necessary to explain why and when the NLOS affects the accuracy. The UWB signal propagation can be used to explain this.

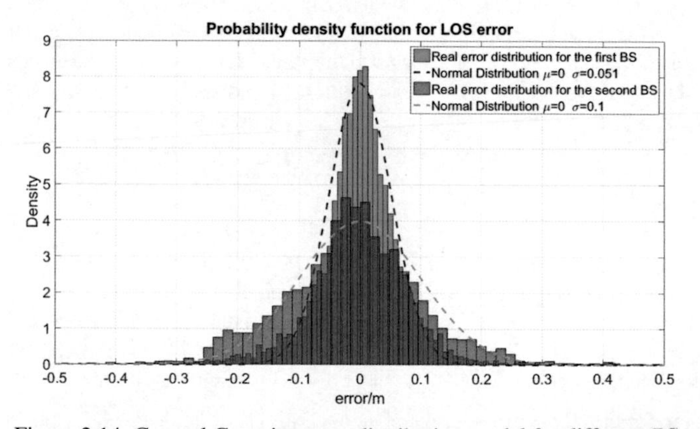

Figure 2.14: General Gaussian error distribution model for different BSs

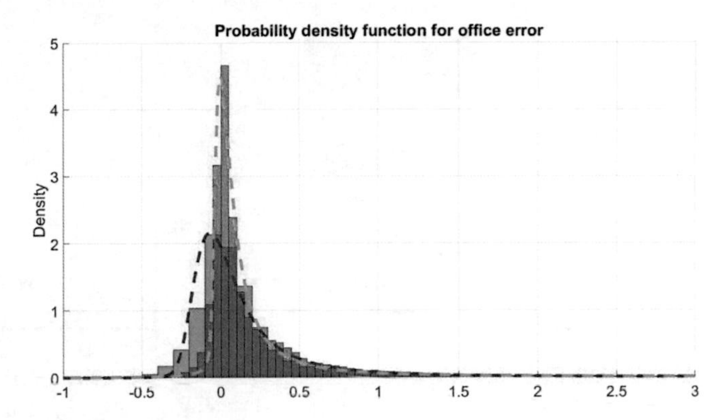

Figure 2.15: Error distribution by combining the LOS and NLOS error for different BSs

2.3 UWB Signal Propagation Description

After the UWB MS sends an impulse, the BS receives not only the impulse from the direct path, but also the impulses from the reflected, diffracted or even scattered path. A channel is used to describe the signal propagation process between the MS and BS. The CIR $c(t)$ is the sum of all received pulses. The signal propagation process is illustrated in Fig. 2.16. The CIR can be described with the following equation:

$$c(t) = \sum_{k=1}^{K} a_k \delta(t - \tau_k) \tag{2.17}$$

where K is the total number of multipath components, a_k and τ_k represent the amplitude and time delay of the kth arrived path. τ_1 is the arrival time of the first arrived path and $\tau_{strongest}$ is the arrival time of the strongest path. Under most LOS conditions, the first arrived path is the strongest path. However, in NLOS conditions, this path is attenuated and might not be the strongest.

DecaWave chips are used in the developed UWB systems. The UWB impulse of the MS and the CIR received in the BS in a real case is presented in Fig. 2.17. A threshold is used to detect the first arrived path. If the impulse from the direct path is not blocked and can be detected by the threshold, the accurate arrival time can be determined. Based on the time difference between the pulse sending time and the arrival time, the range can be calculated by multiplying the time difference by the speed of light.

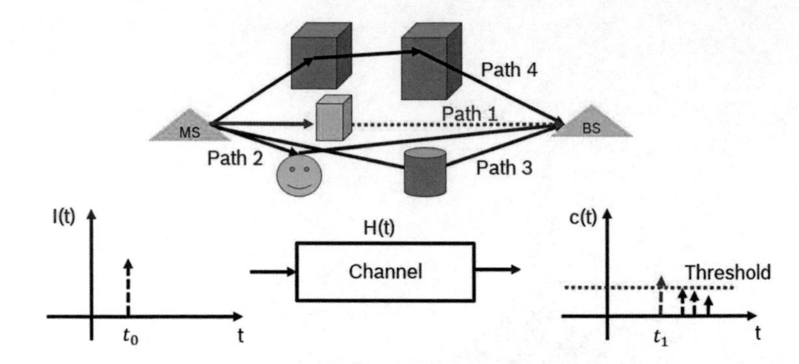

Figure 2.16: Illustration of the signal propagation process

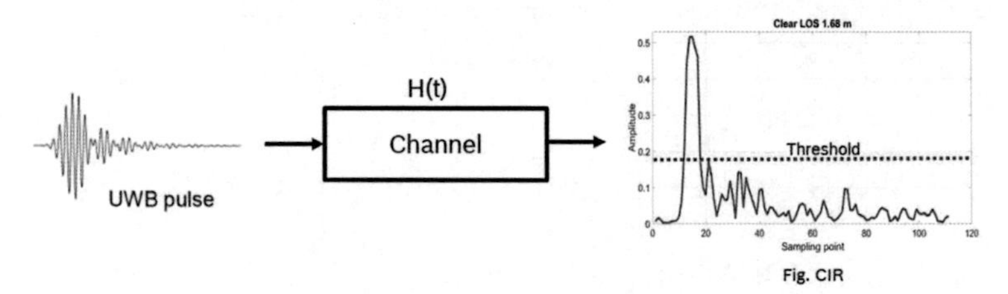

Figure 2.17: Impulse from DecaWave chip and received CIR

2.3.1 Theoretical Explanation for the Relationship between CIRs and Accurate/Inaccurate Measurements

Depending on the environment and the signal propagation path, the measured ranges based on the CIRs can be accurate or inaccurate. The relationship between the CIRs and the accurate/inaccurate ranges is discussed in three different cases: ideal LOS path, small-scale fading: multipath and NLOS path.

Ideal LOS Path

UWB path loss can be defined as the UWB signal power density attenuation when it propagates from the MS to the BS. The path loss in logarithmic units db can be represented by the following equation [LJHV13]:

$$PL(d) = \widetilde{PL_0} + 10\kappa log(\frac{d}{d_0}) + S, d \geq d_0, \tag{2.18}$$

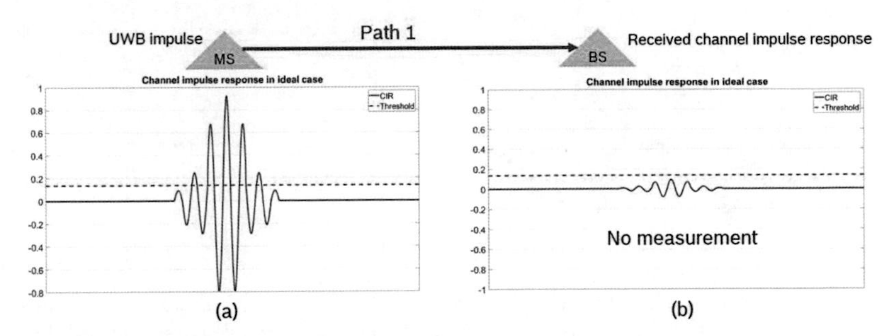

Figure 2.18: (**a**) The received signal can be detected by the threshold; (**b**) The received signal can not be detected by the threshold.

where $\widetilde{PL_0}$ is the mean path loss at the reference distance $d_0 = 1m$; κ is the path loss exponent related to the propagation environment; and S represents the zero mean Gaussian random variable, with standard deviation σ_0, which is used to model the large-scale fading.

Assuming an ideal LOS case, the BS only receives the UWB signal from the direct LOS path. There are two possibilities. First, the received power is large enough that the received signal can be detected by the threshold, as shown in Figure 2.18 (a). In this case, the measured range is accurate. Second the distance between the BS and MS is too large, so the received signal is below the threshold and can not be detected (Figure 2.18 (b)). In this situation, the range can not be measured.

Small-Scale Fading: Multipath

In reality, the BS receives not only the signal from the direct path but also the reflected, diffracted or scattered signals, which is called multipath. These signals reach the BS at different times and might interfere with the received signal from the direct path. In the multipath situation, three different possibilities need to be discussed.

1) The reflected signals have little influence of the first arrived signal.

If the reflected path is much longer than the direct path, the delay of the reflected signal is so large that the reflected signals will not interfere with the signal from the direct path, as shown in Figure 2.19.

2) The reflected signal influences the first arrived signal.

If the reflected signal and the signal from the direct path arrive in the BS at almost the same time, the threshold might not be able to detect the sum of these two signals. The

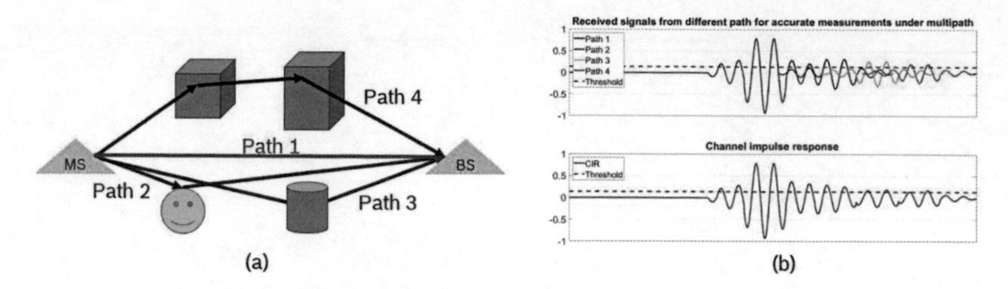

Figure 2.19: (a) Description of the reflected signals and the signal from the direct path. (b) Received CIR with no interference of the reflected signals and the signal from the direct path.

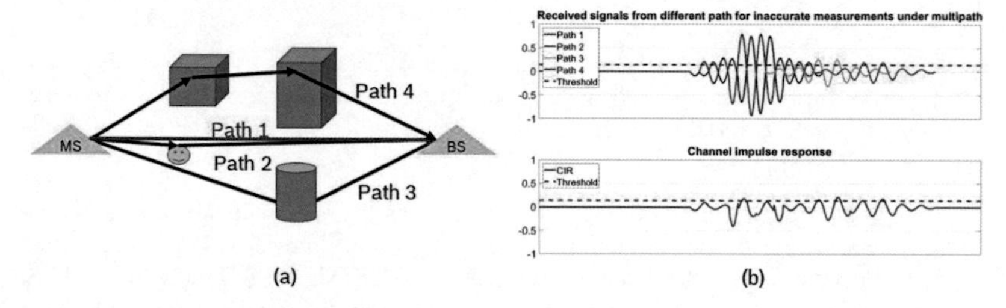

Figure 2.20: (a) Description of the interference of the reflected signals and the signal from the direct path. (b) Received CIR with the interference of the reflected signals and the signal from the direct path.

measured ranges under this condition are not accurate, as shown in Figure 2.20.

3) The distance between the BS and MS is too large.

If the distance is so large that the revived power is too small, the measurement of the range is not possible.

NLOS

If the signal propagation between the BS and the MS is blocked and the received power is so large that the threshold is still able to detect the signals (as shown in Figure 2.21), then theoretically the NLOS error can be calculated using the following equations [Deca]:

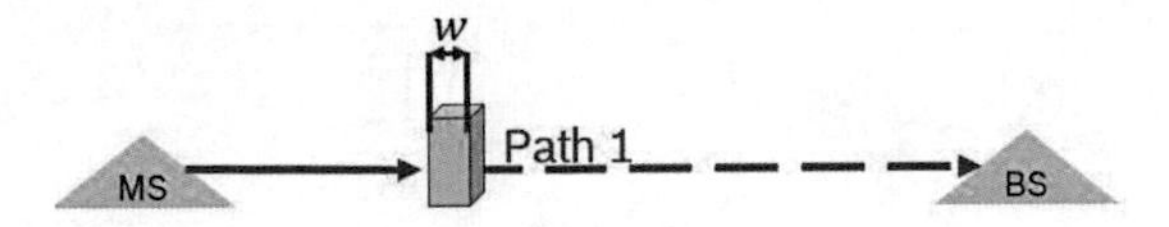

Figure 2.21: weak NLOS blockage

$$T = \frac{d-w}{c} + \frac{wR}{c} = \frac{d+w(R-1)}{c} \tag{2.19}$$

$$\widetilde{d} = Tc = d + w(R-1) \tag{2.20}$$

$$error = \widetilde{d} - d = w(R-1) \tag{2.21}$$

where T is the signal propagation time between BS and MS; d is the real distance between the BS and MS, while $\widetilde{d}$ is the measured distance between them; c is the speed of light; w is the width of the obstruction; R is the refractive index of the obstruction; and error is the theoretical NLOS error. DecaWave provides the refractive indices of some materials, as shown in Table 2.2 [Deca].

Table 2.2: Refractive indices of various materials [Deca]

Material	**Dielectric Constant**	**Refractive Index**
Concrete-solid	7.5	2.73
Concrete-hollow	3	1.73
Dry Wall	2	1.41
Human body	51	7.14
Plywood	2	1.41
Water	80	8.94 (at 4 GHz)

NLOS is almost always combined with multipath, so it always needs to be considered with multipath. Four possibilities need to be taken into account under the NLOS condition.

1) The obstruction can be omitted.

If the signal propagation path is blocked by very thin glass or books, the power reduction and signal arrival delay are so small that accurate range measurements can still be obtained, as shown in Figure 2.22.

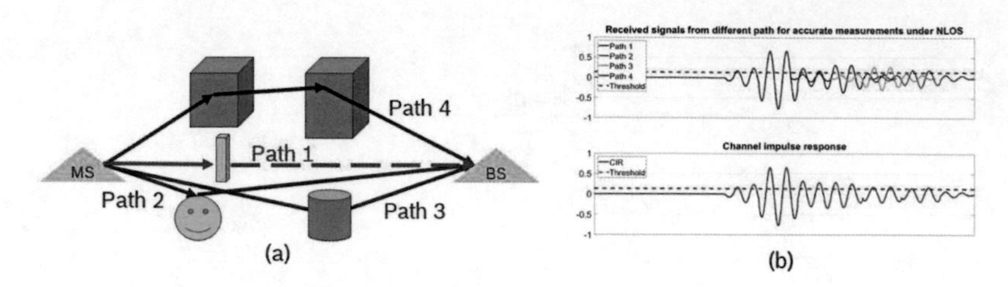

Figure 2.22: (**a**) Description of ignorable blockages. (**b**) Received CIR with ignorable blockages.

2) The signal from the direct path can still be detected with a proper threshold, but the delay causes inaccurate measurements

Sometimes, even with the blockage between the MS and BS, the signal from the direct path can still be detected. However, due to the arrival delay, the measured ranges are inaccurate, as shown in Figure 2.23.

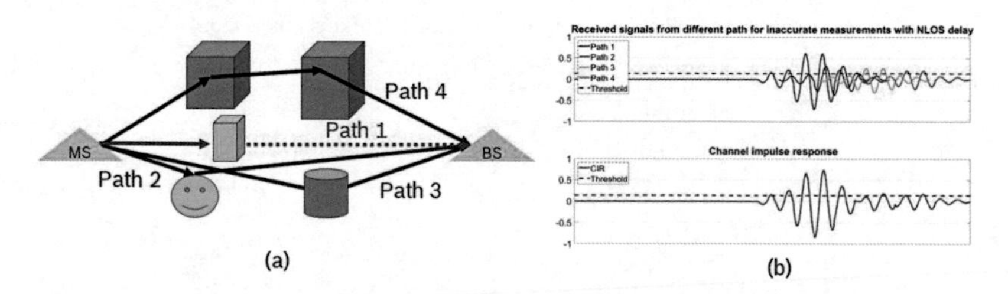

Figure 2.23: (**a**) Description of non-ignorable blockages. (**b**) Received CIR with non-ignorable blockages.

3) If the signal from the direct path is totally blocked, the reflected signals are detected.

If the direct signal propagation path is blocked by thick metal, this signal can not be received by the BS. The reflected signals arrive at the BS with delay and are detected by the threshold, as shown in Figure 2.24. A positive bias is added to the range measurements in this case.

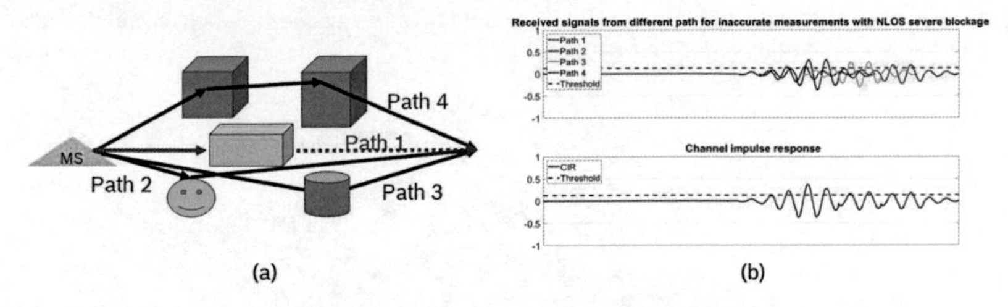

Figure 2.24: (**a**) Description of total blockage of the signal from the direct path. (**b**) Received CIR with total blockage of the signal from the direct path.

4) The distance between the BS and MS is too large.

No measurements can be obtained if the distance between the BS and MS is too large.

In summary, accurate and inaccurate measurements are respectively obtained in the following situations.

Accurate measurement: 1. clear LOS; 2. multipath without interference of the first arrived path; and 3. detected first path with ignorable delay of the NLOS blockage.

Inaccurate measurement: 1. multipath with strong interference of the first arrived path; 2. NLOS blockage with non-ignorable delay of the first path; and 3. multipath and total NLOS blockage of the first path.

The situation in which the distance between the BS and MS is too large is not important in this paper, because no range measurements can easily be detected and will not cause inaccurate position estimation.

2.3.2 Support Vector Machine

The CIRs for accurate/inaccurate measurements are obtained in different situations and they have different characteristics; for example, the energy is different at the same distance. These different characteristics can be used as features in machine learning algorithms to differentiate between the CIRs for accurate and inaccurate measurements.

Machine learning can be divided into three types: supervised, unsupervised and reinforcement learning. Supervised learning deals with problems, in which labels are available, such as classification, regression, and anomaly detection. Unsupervised learning can be used to solve clustering problems without labels. In robotics, where feedback is usually available, reinforcement learning is very popular. In NLOS detection, the CIRs are labeled as CIRs for accurate ranges and NLOS CIRs. This is a typical two-class classification supervised learning problem. In the ideal case, if the extracted features can be totally separated, 100 % of the CIRs for inaccurate measurements can be detected, as indicated in Figure 2.25.

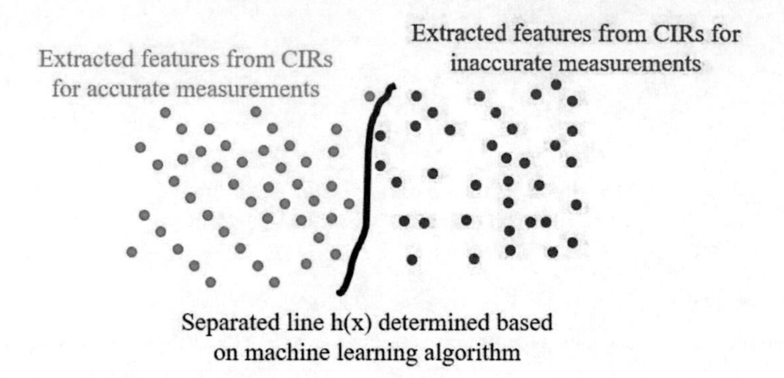

Figure 2.25: Separation of CIRs for accurate and inaccurate measurements in the ideal case

Many characteristic differences between the CIRs for accurate and inaccurate measurements can be observed. These differences can be used as features. For example, at the same distance, the difference between the LOS and NLOS can be more easily observed based on energy and maximal amplitude. Due to the blockage, the energy and maximal amplitude of the CIRs under NLOS are lower compared to those under LOS. Due to the delay of the first path or the reflection of the blockage etc., the shape of the CIR changes a lot. The features that can be used to describe CIRs' shape, can be used to identify the NLOS CIRs. Due to the delay or the attenuation of the blocked arrival signals, the time to the maximal amplitude is longer. It could even happen that the maximal amplitude is not the first path but the arrived multipath. With a blockage, the multipath usually happens more frequently. The received power is also different under NLOS and LOS conditions. Thus, any features that can be used to describe these differences can be used for NLOS identification. With the help of these characteristics, these features can be extracted from five aspects: distance based; shape based; time based; multipath richness based; and power based factors. Details about feature extraction can be found in Chapter 5 . For different feature combinations, the identification accuracy differs. Thus, the optimal feature combination must be determined.

Algorithms such as decision forest, decision jungle, boosted decision tree, neural network, and support vector machine (SVM) can be used to solve classification problems. Compared to other algorithms, SVM is more suitable for the determination of the optimal feature combination, since it provides the same solutions for the problem with the same classifier configuration. However, the solution of the multilayer neural network can be various local minima.

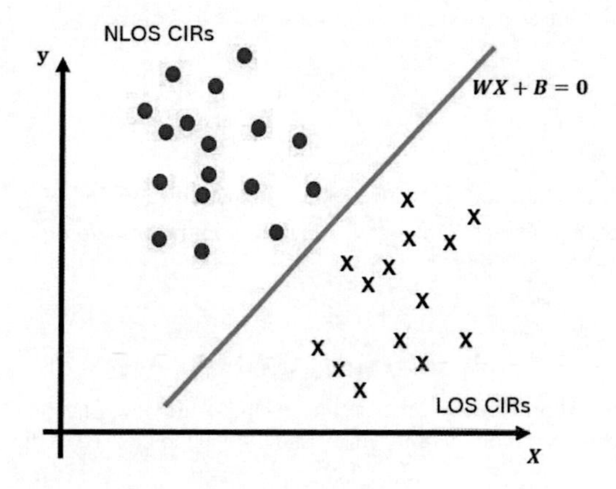

Figure 2.26: Separation of CIRs for accurate and inaccurate measurements in the ideal case

SVM can be used to solve linear and non-linear classification problems. To explain the basic principle of SVM, assume that we have a line classification problem. As shown in Figure 2.26, the red circles represent the NLOS CIRs, and the blue crosses represent the LOS CIRs. SVM is used to find the hyperplane with the largest margin to separate the NLOS CIRs and LOS CIRs. Assume the training dataset T and the hyperplane can be represented as:

$$T = \{(x_1, y_1), (x_2, y_2), \dots, (x_N, y_N)\} \tag{2.22}$$

$$wx + b = 0 \tag{2.23}$$

The geometric margin from the i point to the hyperplane can be calculated with the following equation:

$$\gamma_i = y_i(\frac{w}{||w||}x_i + \frac{b}{||w||}) \tag{2.24}$$

The smallest geometric margin of all the points is:

$$\gamma = \min_{i=1,2,\dots,N} \gamma_i \tag{2.25}$$

The hyperplane with the largest margin can be calculated by solving the following optimization problem:

$$\max_{w,b} \gamma$$

$$s.t.\ y_i(\frac{w}{||w||}x_i + \frac{b}{||w||}) \geq \gamma, i = 1, 2, \ldots, N \tag{2.26}$$

Since $||w||$ and γ are scalars, for simplification, we can assume:

$$w = \frac{w}{||w||\gamma} \tag{2.27}$$

$$b = \frac{b}{||w||\gamma} \tag{2.28}$$

Since the maximization of γ is equivalent to the maximization of $\frac{1}{||w||}$, which is also equivalent to the minimization of $\frac{1}{2}||w||^2$, thus, the optimization problem can be simplified as:

$$\begin{aligned} &\min_{w,b} \frac{1}{2}||w||^2 \\ &s.t.\ y_i(wx_i + b) \geq 1, i = 1, 2, \cdots, N \end{aligned} \tag{2.29}$$

If the optimization problem can not be linear separated, which happens most of the time in the reality, two more parameters (ξ_i and C) need to be used in the optimization problem:

$$\begin{aligned} &\min_{w,b,\xi} \frac{1}{2}||w||^2 + C\sum_{i=1}^{n} \xi_i \\ &s.t.\ y_i(wx_i + b) \geq 1 - \xi_i, \\ &\xi_i \geq 0, i = 1, 2, \cdots, N \end{aligned} \tag{2.30}$$

where ξ_i is the slack variable that is used to tolerate misclassification in case the ideal separation of the classes is not possible. C is the regularization parameter to control the toleration level of the misclassification in the training example.

For non-linear classification, a kernel function must be used. The Gaussian radial basis function kernel is used to solve the classification problem in this thesis. There are two parameters to be tuned to improve the classification accuracy: C and Gamma. An improper C and Gamma could lead to overfitting or inaccurate classification problems. More details about the SVM algorithm can be found in [BWH11], [Kec01]. The training dataset is built by the selected feature combination and classified into two groups: the NLOS CIR and the CIR for accurate range groups. Based on the SVM algorithm and the training dataset, the SVM model can be built and used for classification.

2.3.3 CIR and Accurate/Inaccurate Measurements in Real Office Environments

To evaluate the relationship between the CIRs and the accurate/inaccurate measurements, more than 23,319 CIRs are collected in different places under different conditions in the Bosch Shanghai office environment. It is found that accurate range measurements are obtained under three conditions: clear LOS, where there are not many objects around; no blockage under multipath; and ignorable blockage, such as glass, chairs, tables. In contrast, the inaccurate ranges are measured with people, water, metal block and so forth. Figure 2.27 shows some typical CIRs for accurate and inaccurate ranges. Figure 2.27 (a) through (f) are measured in the environments shown in Figure 2.28 (a) through (f). Figure 2.27 (a), (b) and (c) show the CIRs for accurate ranges in three different situations as previously mentioned. While Figure 2.27 (d), (e) and (f) present the typical CIRs for inaccurate ranges. As shown in these figures, the energy and the shape of the CIRs for accurate and inaccurate measurements are different. Based on these differences, the machine learning algorithms can be used to identify inaccurate measurements.

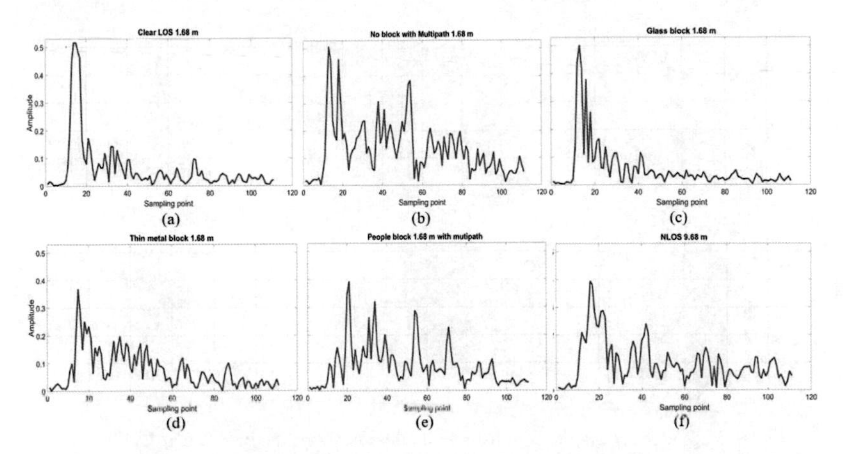

Figure 2.27: CIRs under different environments: (**a**) under clear LOS; (**b**) in multipath without blockage; (**c**) with a glass blockage; (**d**) with a thin metal blockage; (**e**) with a people blockage; (**f**) under NLOS.

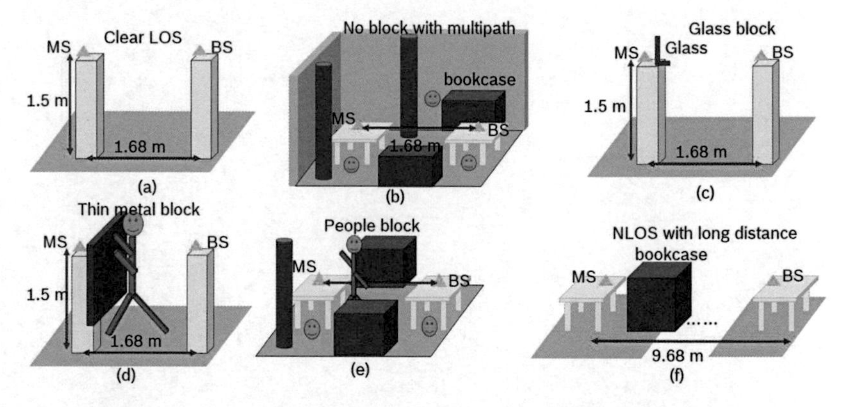

Figure 2.28: Data collection environments description: (**a**) under clear LOS; (**b**) under multipath without blockage; (**c**) with a glass blockage. (**d**) with a thin metal blockage; (**e**) with a people blockage; (**f**) under NLOS.

3 UWB Localization Filter Algorithms

The dynamic UWB localization problem can be described with the following system dynamic models:

$$X_t = f(X_{t-1}, u_{t-1}, w_{t-1}) \tag{3.1}$$

$$Z_t = h(X_t, \xi_t) \tag{3.2}$$

where X is the system state vector, $f(\dots)$ describes the system dynamic and is used to predict the future state vector, u is the input vector, w is the system noise, Z is the measurement vector, $h(\dots)$ is the measurement function, and ξ is the measurement noise. The index t and $t-1$ indicate the time sample.

The following parameters can be predefined or measured in the process:

-The initial state vector X_0, and the initial probability density function $p(X_0)$ for X_0

-Function $f(\dots)$ and $h(\dots)$

-System noise and measurement noise

-The input vector, u_0,u_1,...,u_{t-1}

-The measurement vector, Z_1,Z_2,...,Z_t

The goal is to obtain the best estimated current location of the MS in other words, to calculate the best estimated X_t based on the given parameters and the system dynamic models 3.1 and 3.2. From the Bayesian viewpoint, the best estimated X_t can be obtained with the conditional probability density function (pdf):

$$p(X_t|Z_1, Z_2, \dots, Z_t, \dots, u_0, u_1, \dots, u_{t-1}) \tag{3.3}$$

The system dynamic process can be described as follows:

► Given the initial state vector X_0, initial probability density function $p(X_0)$ for X_0:

-The state vector is predicted based on the system dynamic model 3.1.

-Z_1 is measured.

-The best estimated state vector X_1 and the conditional pdf $p(X_1|Z_1, u_0)$ are updated based on the predicted state vector and the measurements Z_1.

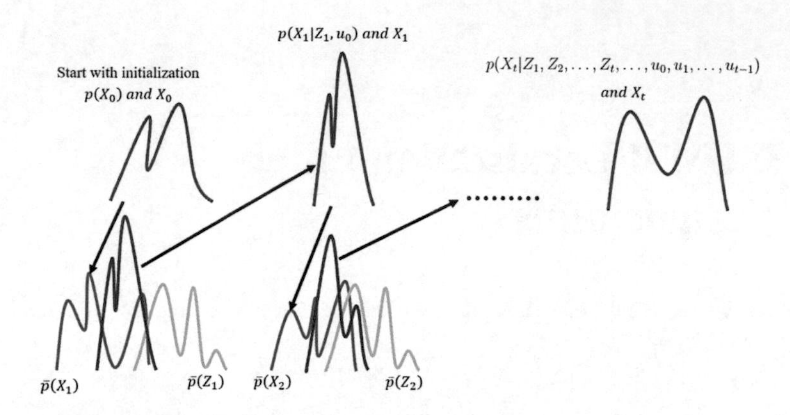

Figure 3.1: System dynamic process explained with conditional probability density

► With the best estimated X_1 state vector and the conditional pdf $p(X_1|Z_1, u_0)$:

-The state vector is predicted based on the system dynamic model 3.1.

-Z_2 is measured.

-The best estimated state vector X_2 and the conditional pdf $p(X_2|Z_1, Z_2, u_0, u_1)$ are updated based on the predicted state vector and the measurements Z_2.

► ⋮

► With the conditional pdf $p(X_{t-1}|Z_1, Z_2, \ldots, Z_{t-1}, \ldots, u_0, u_1, \ldots, u_{t-2})$ and the best estimated X_{t-1} state vector:

-The state vector is predicted based on the system dynamic model 3.1.

-Z_t is measured.

-The conditional pdf $p(X_t|Z_1, Z_2, \ldots, Z_t, \ldots, u_0, u_1, \ldots, u_{t-1})$ and the best estimated state vector X_t are updated based on the predicted state vector and the measurements Z_t.

The system dynamic process can also be described solely with the help of the conditional probability density function (pdf) as shown in Figure 3.1. The initial state vector X_0 and initial probability density function $p(X_0)$ are determined during the initialization. After the first prediction for the state vector based on the system dynamic model 3.1, the pdf for the predicted state vector $\bar{p}(X_1)$ can be obtained. Different algorithms can be used to combine the measurements pdf $\bar{p}(Z_1)$ and $\bar{p}(X_1)$ to compute the best estimated conditional pdf for the state vector $p(X_1|Z_1, u_0)$ and the state vector X_1.

These best estimated parameters are used for the prediction of $\bar{p}(X_2)$. The process is the same as before. Then, the best estimated conditional pdf propagates until time t.

In summary, the following steps are needed to prepare the filter:

1) The input vector u_0 and system state vector need to be determined. Specifically, for UWB based localization, the system state vector and control vector are:

$$X = [x \; y \; v_x \; v_y]' \tag{3.4}$$

$$u = [a_x \; a_y]' \tag{3.5}$$

where (x, y) is the position of the MS and (v_x, v_y) is the velocity of the MS. (a_x, a_y) is the acceleration in x and y directions in global coordinates.

2) The measurement vector Z_1,Z_2,...,Z_t must be determined. For TOA, the ranges are the measurements, while for TDOA the range differences are the measurements.

3) The functions $f(\dots)$ and $h(\dots)$ need to be defined. The function $f(\dots)$ is a linear function and is the same for TOA and TDOA:

$$f(\dots): \;\; X_t = AX_{t-1} + Bu + W \tag{3.6}$$

where W is the system noise vector and

$$A = \begin{pmatrix} 1 & 0 & \Delta t & 0 \\ 0 & 1 & 0 & \Delta t \\ 0 & 0 & 1 & 0 \\ 0 & 0 & 0 & 1 \end{pmatrix} \tag{3.7}$$

$$B = \begin{pmatrix} \frac{1}{2}\Delta t^2 & 0 \\ 0 & \frac{1}{2}\Delta t^2 \\ \Delta t & 0 \\ 0 & \Delta t \end{pmatrix} \tag{3.8}$$

where Δt is the difference between the current time t and the previous time $t-1$.

However $h(\dots)$ is a non-linear function:

$$h(\dots): \;\; Z_t = \begin{pmatrix} \hat{z}_{1,t} \\ \hat{z}_{2,t} \\ \vdots \\ \hat{z}_{N,t} \end{pmatrix} \tag{3.9}$$

where for TOA:

$$\hat{z}_{i,t} = \sqrt{(x_i - x_t)^2 + (y_i - y_t)^2 + (z_i - z_t)^2} \tag{3.10}$$

and for TDOA:

$$\begin{aligned}\hat{z}_{ij,t} &= \sqrt{(x_i - x_t)^2 + (y_i - y_t)^2 + (z_i - z_t)^2} \\ &- \sqrt{(x_j - x_t)^2 + (y_j - y_t)^2 + (z_j - z_t)^2}\end{aligned} \tag{3.11}$$

The position of the ith BS is represented by $X_i = (x_i, y_i, z_i)$, and $X_s = (x_t, y_t, z_t)$ is the real position of the MS. z_i and z_t are constant, since the heights of the MS and BS are fixed.

4) The system noise and measurement noise models must be built. Different filter algorithms, such as Kalman filter and particle filter, are selected based on the built models. The conditional probability density propagation is realized in the algorithms. The system noise and measurements noise models are described in Chapter 2.

3.1 Kalman Filter

The KF is known as a linear quadratic estimator and can be used to propagate the Gaussian noise distribution. The algorithm contains two steps: the prediction step and the update step. In the first step, the estimated current state vector is produced based on the previous state vector and the uncertainty is also calculated. The measurements with the random noise for the current time are observed. In the second step, the estimated current state vector is updated using the calculated weighted average, which is called the Kalman gain. Based on the estimated uncertainties for the estimated current state and the measurements, more weight is given to the vector with higher certainty [ZLW18a], [LLJD15]. To further understand the KF, the following concepts must be classified.

Main Diagonal and Trace

The main diagonal of a matrix A is a collection of $a_{i,j}$, where $i = j$ [mai].

$$A = \begin{pmatrix} a_{11} & a_{12} & \dots & a_{1n} \\ a_{21} & a_{22} & \dots & a_{2n} \\ \vdots & \vdots & \vdots & \vdots \\ a_{n1} & a_{n2} & \dots & a_{nn} \end{pmatrix} \tag{3.12}$$

The trace of matrix A $(tr(A))$ is the sum of the main diagonal:

$$tr(A) = \sum_{i=1}^{n} a_{ii} = a_{11} + a_{22} + \cdots + a_{nn} \tag{3.13}$$

Variance

The variance of a random variable X is defined as follows [Pol]:

$$var(X) = [E(X - \mu_X)^2] \tag{3.14}$$

where $E[Y]$ means the expected value of Y. μ_X is the expected value of X.

Covariance

The covariance between two random variables X and Y is defined as follows [Pol]:

$$cov(X, Y) = [E(X - \mu_X)(Y - \mu_Y)] \tag{3.15}$$

Assuming that X is a column vector and X_i is random variables with μ_i as the expected value, the covariance matrix $cov(X_i, X_j)$ is defined as follows [cov]:

$$X = \begin{pmatrix} X_1 \\ X_2 \\ \vdots \\ X_n \end{pmatrix} \tag{3.16}$$

$$cov(X_i, X_j) = [E(X_i - \mu_i)(X_j - \mu_j)] =$$
$$\begin{pmatrix} E(X_1 - \mu_1)(X_1 - \mu_1) & E(X_1 - \mu_1)(X_2 - \mu_2) & \dots & E(X_1 - \mu_1)(X_n - \mu_n) \\ E(X_2 - \mu_2)(X_1 - \mu_1) & E(X_2 - \mu_2)(X_2 - \mu_2) & \dots & E(X_2 - \mu_2)(X_n - \mu_n) \\ \vdots & \vdots & \vdots & \vdots \\ E(X_n - \mu_n)(X_1 - \mu_1) & E(X_n - \mu_n)(X_2 - \mu_2) & \dots & E(X_n - \mu_n)(X_n - \mu_n) \end{pmatrix} \tag{3.17}$$

The main diagonal of the covariance matrix is the variances.

3.1.1 Kalman Filter Principle

As described above, the KF is a linear quadratic estimator. The main idea of the KF is to define an optimal Kalman gain to combine the predicted estimated state and the measurements. Assume that the linear dynamic model is as follows:

$$X_{r,t} = AX_{r,t-1} + Bu_{t-1} + w_{t-1} \tag{3.18}$$

$$Z_t = HX_{r,t} + \xi_t \tag{3.19}$$

where $X_{r,t}$ is the true state vector. A is the state transition matrix, B is called the control matrix, and w_{t-1} is the system process noise, which has to be a zero mean Gaussian distribution with Q as covariance. Z_t is the measurements. ξ_t is the measurement noise, which has to be a zero mean Gaussian distribution with R as covariance.

Assuming that X_t is the optimized state vector, $\hat{X}_t$ is the predicted vector state and $\hat{Z}_t$ is the predicted measurement vector. The main idea of the KF is to find the optimal K_t to calculate the X_t:

$$X_t = \hat{X}_t + K_t(Z_t - \hat{Z}_t) \tag{3.20}$$

where K_t is the Kalman gain and

$$\hat{X}_t = AX_{t-1} + Bu_{t-1} \tag{3.21}$$

$$\hat{Z}_t = H\hat{X}_t \tag{3.22}$$

To determine K_t, the estimated error covariance P_t between the optimized estimated state vector X_t and the true state vector $X_{r,t}$ must be checked.

$$P_t = E[(X_{r,t} - X_t)(X_{r,t} - X_t)^T] \tag{3.23}$$

Combining equation 3.23 with equations 3.20, 3.22 and 3.19 yields:

$$\begin{aligned} P_t &= E[((I - K_tH)(X_{r,t} - \hat{X}_t) - K_t\xi_t)((I - K_tH)(X_{r,t} - \hat{X}_t) - K_t\xi_t)^T] \\ &= (I - K_tH)E[(X_{r,t} - \hat{X}_t)(X_{r,t} - \hat{X}_t)^T](I - K_tH) + K_tE[\xi_t\xi_t^T]K_t^T \end{aligned} \tag{3.24}$$

Assuming that the estimated error covariance between the predicated vector state $\hat{X}_t$ and the true state vector $X_{r,t}$ is $\bar{P}_t$:

$$\begin{aligned} \bar{P}_t &= E[(X_{r,t} - \hat{X}_t)(X_{r,t} - \hat{X}_t)^T] \\ &= E[(A(X_{r,t-1} - X_{t-1}) + w_{t-1})(A(X_{r,t-1} - X_{t-1}) + w_{t-1})^T] \\ &= E[(A(X_{r,t-1} - X_{t-1}))(A(X_{r,t-1} - X_{t-1}))^T] + E[w_{t-1}w_{t-1}^T] \\ &= AP_{t-1}A^T + Q \end{aligned} \tag{3.25}$$

Thus,

$$\begin{aligned} P_t &= (I - K_tH)\bar{P}_t(I - K_tH) + K_tE[\xi_t\xi_t^T]K_t^T \\ &= \bar{P}_t - K_tH\bar{P}_t - \bar{P}_tH^TK_t^T + K_t(H\bar{P}_tH^T + R)K_t^T \end{aligned} \tag{3.26}$$

As discussed above, the trace of P_t is the variance. With optimal K_t, the trace of P_t should be minimal. Which means that $tr(P_t)$ should be minimal:

$$tr[P_t] = tr[\bar{P}_t] - tr[K_tH\bar{P}_t] - tr[\bar{P}_tII^TK_t^T] + tr[K_t(H\bar{P}_tH^T + R)K_t^T] \tag{3.27}$$

The minimization make it necessary that $\frac{dtr[P_t]}{dK_t}$ is zero:

$$\frac{dtr[P_t]}{dK_t} = -2(H\bar{P}_t)^T + 2K_t(H\bar{P}_tH^T + R) = 0 \tag{3.28}$$

Thus,

$$K_t = \bar{P}_tH^T(H\bar{P}_tH^T + R_t)^{-1} \tag{3.29}$$

The calculated K_t can be used to compute P_t, combining equations 3.29 and 3.26:

$$P_t = (I - K_tH)\bar{P}_t \tag{3.30}$$

3.1.2 Kalman Filter for Localization

Based on the above discussion, the KF for the localization problem can be summarized as follows. First, the state vector X and the control vector u need to be determined:

$$X = [x\ y\ v_x\ v_y]' \tag{3.4}$$

$$u = [a_x\ a_y]' \tag{3.5}$$

where (x, y) is the position of the MS and (v_x, v_y) is the velocity of the MS. (a_x, a_y) is the acceleration in x and y directions in global coordinates.

The acceleration is used to predict the state vector for the next time stamp [ZLW18a]. The state predictive transition model can be represented as:

$$\hat{X}_t = AX_{t-1} + Bu_{t-1} \tag{3.21}$$

$$\bar{P}_t = AP_{t-1}A^T + Q \tag{3.25}$$

Q is the process noise covariance matrix.

The measurement model can be obtained with the following equation:

$$\hat{Z}_t = \begin{pmatrix} \hat{z}_{1,t} \\ \hat{z}_{2,t} \\ \vdots \\ \hat{z}_{N,t} \end{pmatrix} = H\hat{X}_t \tag{3.31}$$

where $\hat{Z}_t$ is the predicted measurements, and N is the number of used BSs.

The update process can be realized with the following equations:

$$K_t = \bar{P}_tH^T(H\bar{P}_tH^T + R_t)^{-1} \tag{3.29}$$

$$X_t = \hat{X}_t + K_t(Z_t - \hat{Z}_t) \tag{3.20}$$

$$P_t = (I - K_t H)\bar{P}_t \tag{3.30}$$

where Z_t is the real measurement, K_t is the Kalman gain and R_t is the measurement noise covariance matrix. If the measurement noises are constant, R_t can be written as:

$$R_t = \begin{pmatrix} \xi_1\xi_1 & 0 & 0 & \dots & 0 \\ 0 & \xi_2\xi_2 & 0 & \dots & 0 \\ \vdots & \vdots & \vdots & \dots & \vdots \\ 0 & 0 & 0 & \dots & \xi_N\xi_N \end{pmatrix} \tag{3.32}$$

where ξ_i denotes the standard deviation of the measurement noise for the i_{th} BS under LOS .

To obtain the optimal solution in the KF, the noise has to be Gaussian. The reason for this can be explained with the help of Figure 3.2: assume that the error distributions for the predicted value and the measurements are Gaussian distributions and the standard deviations are σ_1, σ_2. After the update process in the KF, the calculated error distribution is still a Gaussian distribution, but with a small standard deviation (σ_t). This distribution is used for further calculation. Based on this system characteristic, the noise in the KF is kept as a Gaussian distribution. After the update process, the updated results are optimized with a smaller standard deviation. If the noise is not Gaussian, the Gaussian distribution propagation is not possible and the estimated state vector with the smallest variance can not be guaranteed. A structured flowchart for the KF is presented in Figure 3.3.

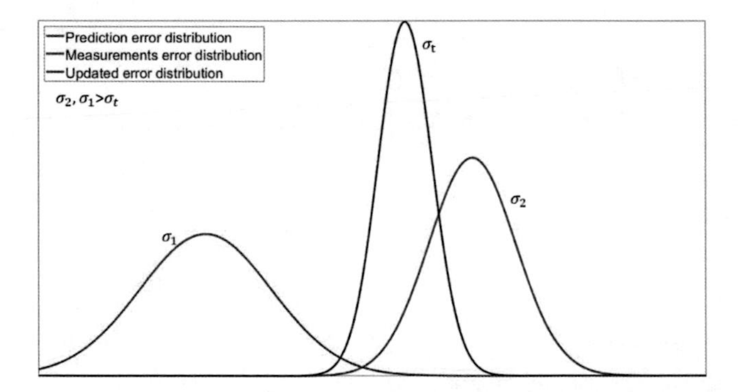

Figure 3.2: Updated noise distribution in the KF

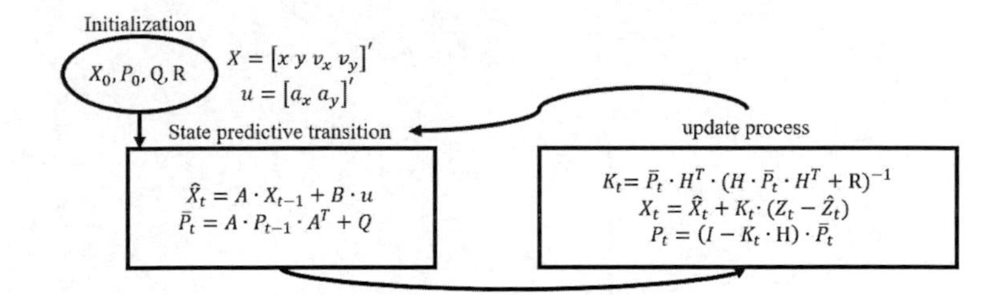

Figure 3.3: Structured flowchart for the KF

3.2 Extended Kalman Filter

The KF is used to solve linear optimization problems. However, for UWB localization, while the state predictive transition is linear, the updated observation model is non-linear. The predicted measurements are calculated using the following equations.

For TOA:

$$\hat{z}_{i,t} = \sqrt{(x_i - x_t)^2 + (y_i - y_t)^2 + (z_i - z_t)^2} \tag{3.33}$$

For TDOA:

$$\begin{aligned} \hat{z}_{ij,t} = & \sqrt{(x_i - x_t)^2 + (y_i - y_t)^2 + (z_i - z_t)^2} \\ & - \sqrt{(x_j - x_t)^2 + (y_j - y_t)^2 + (z_j - z_t)^2} \end{aligned} \tag{3.34}$$

The measurement model is represented by the following equation:

$$\hat{Z}_t = h(\hat{X}_t) \tag{3.35}$$

The extended Kalman filter (EKF) is the non-linear version of the KF, and it is used to solve non-linear problems [MMK09]. In the EKF, non-linear equations need to be linearized. During the linearization, the transformation matrix for TOA is calculated as:

$$H = \frac{\partial h}{\partial X} = \begin{pmatrix} \frac{\partial \hat{z}_{1,t}}{\partial x} & \frac{\partial \hat{z}_{1,t}}{\partial y} & 0 & 0 \\ \frac{\partial \hat{z}_{2,t}}{\partial x} & \frac{\partial \hat{z}_{2,t}}{\partial y} & 0 & 0 \\ \vdots & \vdots & \vdots & \vdots \\ \frac{\partial \hat{z}_{N,t}}{\partial x} & \frac{\partial \hat{z}_{N,t}}{\partial y} & 0 & 0 \end{pmatrix} \tag{3.36}$$

$$\frac{\partial \hat{z}_{i,t}}{\partial x} = \frac{x_i - x_t}{\sqrt{(x_i - x_t)^2 + (y_i - y_t)^2 + (z_i - z_t)^2}} \tag{3.37}$$

$$\frac{\partial \hat{z}_{i,t}}{\partial y} = \frac{y_i - y_t}{\sqrt{(x_i - x_t)^2 + (y_i - y_t)^2 + (z_i - z_t)^2}} \tag{3.38}$$

For TDOA:

$$H = \frac{\partial h}{\partial X} = \begin{pmatrix} \frac{\partial \hat{z}_{12,t}}{\partial x} & \frac{\partial \hat{z}_{12,t}}{\partial y} & 0 & 0 \\ \frac{\partial \hat{z}_{13,t}}{\partial x} & \frac{\partial \hat{z}_{13,t}}{\partial y} & 0 & 0 \\ \vdots & \vdots & \vdots & \vdots \\ \frac{\partial \hat{z}_{1N,t}}{\partial x} & \frac{\partial \hat{z}_{1N,t}}{\partial y} & 0 & 0 \\ \vdots & \vdots & \vdots & \vdots \\ \frac{\partial \hat{z}_{N-1N,t}}{\partial x} & \frac{\partial \hat{z}_{N-1N,t}}{\partial y} & 0 & 0 \end{pmatrix} \tag{3.39}$$

$$\begin{aligned} \frac{\partial \hat{z}_{ij,t}}{\partial x} = & \frac{x_i - x_t}{\sqrt{(x_i - x_t)^2 + (y_i - y_t)^2 + (z_i - z_t)^2}} \\ & - \frac{x_j - x_t}{\sqrt{(x_j - x_t)^2 + (y_j - y_t)^2 + (z_j - z_t)^2}} \end{aligned} \tag{3.40}$$

$$\begin{aligned} \frac{\partial \hat{z}_{ij,t}}{\partial y} = & \frac{y_i - y_t}{\sqrt{(x_i - x_t)^2 + (y_i - y_t)^2 + (z_i - z_t)^2}} \\ & - \frac{y_j - y_t}{\sqrt{(x_j - x_t)^2 + (y_j - y_t)^2 + (z_j - z_t)^2}} \end{aligned} \tag{3.41}$$

After the linearization, the EKF follows similar steps as the KF. Compared to the KF, the optimal solution with minimized variance in the EKF cannot always be guaranteed. Based on the factor that only an approximation can be given for the non-linear model after linearization, the EKF provides only an approximated optimal solution. Similar to the KF, the noise model needs to be Gaussian. However, due to the non-linearity, the propagated conditional probability density function cannot be guaranteed to be Gaussian. Furthermore, if the linearization is not a good approximation for the non-linear equations, the EKF may have a divergence problem.

As the MS moves from LOS to NLOS, the corresponding noise distribution changes from LOS noise distribution to NLOS noise distribution. Once the NLOS measurements are detected, the corresponding ξ_i in R_t must be changed to the NLOS measurement noise. However, as shown in Chapter 2, the standard deviation of the NLOS noise can change dramatically if the blockage changes. A general model for the NLOS error is highly difficult to build. Thus, a parameter de_i is used to tune the noise in the measurement noise covariance matrix. R_t can be represented as:

$$R_t = \begin{pmatrix} de_1\xi_1\xi_1 & 0 & 0 & \dots & 0 \\ 0 & de_2\xi_2\xi_2 & 0 & \dots & 0 \\ \vdots & \vdots & \vdots & \dots & \vdots \\ 0 & 0 & 0 & \dots & de_N\xi_N\xi_N \end{pmatrix} \tag{3.42}$$

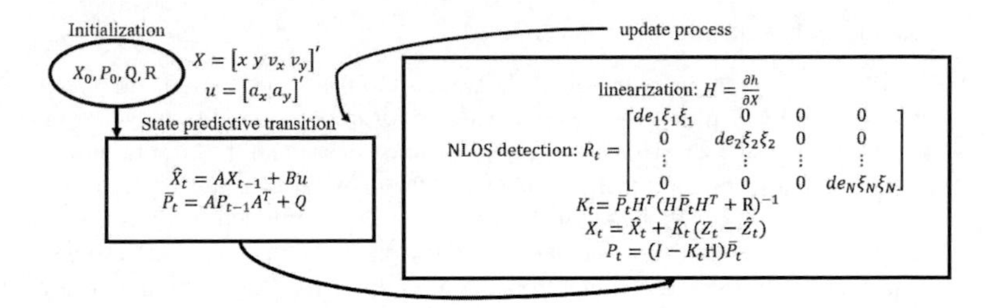

Figure 3.4: Structured flowchart for the KF

If the measurement is detected as accurate, the corresponding de_i is set to be a small value; otherwise, the value is set to be large, so that the influence of the corresponding inaccurate measurement can almost be ignored. A structured flowchart for the EKF with NLOS identification is shown in Figure 3.4.

Furthermore, during the testing, since NLOS happens quite often, the frequent change of the noise in R_t could lead to an instability problem in the EKF. One possible way to improve the stability performance is to use the iterated extended Kalman filter (IEKF).

3.3 Iterated Extended Kalman Filter

Compared to the EKF, the IEKF also contains an iterative process [AKP15]. This process starts at the linearization phase. For j=1,2 ...n (n is the maximum iterative number):

$$H_j = \frac{\partial h}{\partial X_j} \tag{3.43}$$

$$K_{t,j} = \bar{P}_t H_j^T (H_j \bar{P}_t H_j^T + R_t)^{-1} \tag{3.44}$$

$$\hat{X}_{t,j+1} = \hat{X}_t + K_{t,j}(Z_t - h(\hat{X}_{t,j}) - H_j(\hat{X}_t - \hat{X}_{t,j})) \tag{3.45}$$

$$P_{t,j+1} = (I - K_{t,j} H_j)\bar{P}_t \tag{3.46}$$

The condition to end the "for" loop depends on the iterative number n and the absolute difference between $\hat{X}_{t,j+1}$ and $\hat{X}_t$. The remaining steps are the same as in the EKF.

3.4 Particle Filter

Unlike the KF, the particle filter can be used to solve non-linear and non-Gaussian noise problems. The basic idea of particle filter for localization can be described as follows:

1) The initial position of the MS is determined.

2) The MS moves to the next position. Based on the pre-knowledge of the system dynamic, the current position can be predicted. For example, if the MS moves with a constant velocity or the acceleration of the MS can be measured, the current position can predicted. However, the prediction contains error, so particles are set around the predicted position. If the variance of the errors can be determined, the positions of these particles are limited to an area based on the variance of the errors. If absolutely no knowledge about the errors can be obtained, the particles can be assumed to be uniformly distributed across the whole floor. In this case, the used number of particles is very large, and the calculation load is huge.

3) Assuming that the ranges are the measurements, for each particle, a predicted range for each BS can be calculated. The difference between the predicted range and the measured range is used to determine the weight of the particle. The smaller the difference is, the larger the weight is given to this particle.

4) Resampling: The particles with small weights are abandoned. More particles are set to these particles with large weights.

5) Repeat, start from 2) to 4).

The main steps in the particle filter are: setting the particles; calculating the weights of the particles based on the measurements; and resampling. The particle filter is a very powerful filter. To determine its advantages and disadvantages and under which conditions it is the optimal solution, the following discusses the particle filter in further detail.

The particle filter is based on the Monte Carlo method. The main idea is to use the particles to describe the pdf. Since the pdf is approximated with the particles, theoretically with $N \rightarrow \inf$, any kind of pdf can be described. Thus, the PF can deal with linear and non-linear optimization problems with Gaussian or non-Gaussian noise. To understand the PF, the following concepts need to be explained first.

Probability

Probability describes the likelihood that an event will occur. It is limited between 0 and 1, where 0 means that an event is impossible, 1 means that it will certainly happen. $P(A)$ indicates the probability of an event A [pro], [Mer13].

Frequency

Assuming that experiments run N times and the event A happens n_A times, the frequency that event A happens is defined as:

$$f(A) = \frac{n_A}{N} \tag{3.47}$$

If $N \to \inf$, theoretically $f(A)$ can be assumed to be the same as $P(A)$. This is the basis of the Monte Carlo method.

$$f(A)_{N \to \inf} = P(A) \tag{3.48}$$

Conditional Probability

Conditional probability describes the probability of an event happening under the condition that another event has happened. $P(A|B)$ represents the probability of A under the condition of B [con].

Monte Carlo Method

The basic principle of the Monte Carlo method can be explained as follows. Assuming that a distribution $p(x)$ is known, if a set of N samples are drawn independently and are identically distributed, then the distribution $p(x)$ can be approximated with these samples under the condition that $N \to \inf$. The expectation can also be calculated with these samples [Ho]. The basic idea of the PF is based on this method, which means using particles to describe the distributions. The expectation can be obtained as follows [Ter]:

$$E[f(x)] = \int_a^b f(x)p(x)dx \approx \frac{1}{n}\sum_{i=1}^{n} f(x_i)_{n \to \inf} \tag{3.49}$$

Importance Sampling

Assuming that $q(x)$ is the importance density, the expectation of $f(x)$ can be calculated with the following equations based on the Monte Carlo method [Ter]:

$$\begin{aligned} E[f(x)] = \int_a^b f(x)p(x)dx &= \int_a^b \frac{f(x)p(x)}{q(x)} q(x)dx = E[\frac{f(x)p(x)}{q(x)}] \\ &\approx \frac{1}{n}\sum_{i=1}^{n} \frac{f(x_i)p(x_i)}{q(x_i)} = \frac{1}{n}\sum_{i=1}^{n} W(x_i)f(x_i) \end{aligned} \tag{3.50}$$

where $W(x_i)$ is the normalized importance weights and

$$W(x_i) = \frac{p(x_i)}{q(x_i)} \quad and \quad \sum_{i=1}^{n} W(x_i) = 1 \tag{3.51}$$

Bayes' Rule

The conditional probability can be calculated based on Bayes' rule. This theorem can be represented by the following equation [Wes]:

$$P(A|B) = \frac{P(B|A)P(A)}{P(B)} \tag{3.52}$$

Markov Process

A process is defined as a Markov process if the next state is solely determined by the current state, which can be expressed as [WH]:

$$p(X_t|X_{t-1}, X_{t-2}, \ldots, X_1) = p(X_t|X_{t-1}) \tag{3.53}$$

The localization process is a Markov process, since the position at time $t+1$ solely depends on the position at time t. In the following chapters, $X_{t-1}, X_{t-2}, \ldots, X_1$ is written as $X_{1:t-1}$.

3.4.1 Particle Filter Principle

Assume that the following equations are used to describe the system dynamic model. The initial pdf $p(X_0|Z_0)$ is known. $p(X_{t-1}|Z_{1:t-1})$ is determined in the last step. The pdf $p(X_t|Z_{1:t})$ needs to be calculated.

$$X_t = f(X_{t-1}, u_{t-1}, w_{t-1}) \tag{3.54}$$

$$Z_t = h(X_t, \xi_t) \tag{3.55}$$

The calculation of $p(X_t|Z_{1:t})$ can be realized in two steps: a prediction step and an update step.

Prediction Step

Based on the Chapman-Kolmogorov equation, $p(X_t|Z_{1:t-1})$ can be computed with $p(X_{t-1}|Z_{1:t-1})$:

$$\begin{aligned} p(X_t|Z_{1:t-1}) &= \int p(X_t, X_{t-1}|Z_{1:t-1})dX_{t-1} \\ &= \int p(X_t|X_{t-1}, Z_{1:t-1})p(X_{t-1}|Z_{1:t-1})dX_{t-1} \end{aligned} \tag{3.56}$$

Based on the fact that the localization is a Markov process:

$$p(X_t|X_{t-1}, Z_{1:t-1}) = p(X_t|X_{t-1}) \tag{3.57}$$

where $p(X_t|X_{t-1})$ can be obtained based on function $X_t = f(X_{t-1}, u_{t-1}, w_{t-1})$ and w_{t-1}. Thus, $p(X_t|Z_{1:t-1})$ can be calculated with the following equation:

$$p(X_t|Z_{1:t-1}) = \int p(X_t|X_{t-1})p(X_{t-1}|Z_{1:t-1})dX_{t-1} \tag{3.58}$$

Update Step

Based on Bayes' rule, $p(X_t|Z_{1:t})$ can be computed as:

$$p(X_t|Z_{1:t}) = \frac{p(Z_t|X_t)p(X_t|Z_{1:t-1})}{p(Z_t|Z_{1:t-1})} \tag{3.59}$$

where $p(Z_t|X_t)$ can be obtained using the function $Z_t = h(X_t, \xi_t)$ and ξ_t and

$$p(Z_t|Z_{1:t-1}) = \int p(Z_t|X_t)p(X_t|Z_{1:t-1})dX_t \tag{3.60}$$

The calculation of the integration equations in the prediction and update steps for a non-linear and non-Gaussian system is highly difficult. Based on the Monte Carlo method and the importance sampling, the particle filter can calculate the approximated optimal solution for these equations even for a non-linear and non-Gaussian system with the particles. The expectation of $f(X_t)$ can be calculated as follows:

$$\begin{aligned} E[f(X_t)] &= \int f(X_t)p(X_t|Z_{1:t})dX_t \\ &= \int f(X_t)\frac{p(X_t|Z_{1:t})}{q(X_t|Z_{1:t})}q(X_t|Z_{1:t})dX_t \\ &= \int f(X_t)\frac{p(Z_{1:t}|X_t)p(X_t)}{p(Z_{1:t})q(X_t|Z_{1:t})}q(X_t|Z_{1:t})dX_t \\ &= \int f(X_t)\frac{W(X_t)}{p(Z_{1:t})}q(X_t|Z_{1:t})dX_t \end{aligned} \tag{3.61}$$

where $W(X_t)$ is the unnormalized importance weights [Ter].

$$W(X_t) = \frac{p(Z_{1:t}|X_t)p(X_t)}{q(X_t|Z_{1:t})} \propto \frac{p(X_t|Z_{1:t})}{q(X_t|Z_{1:t})} \tag{3.62}$$

Thus

$$\begin{aligned} E[f(X_t)] &= \frac{1}{p(Z_{1:t})}\int f(X_t)W(X_t)q(X_t|Z_{1:t})dX_t \\ &= \frac{\int f(X_t)W(X_t)q(X_t|Z_{1:t})dX_t}{\int p(Z_{1:t}|X_t)p(X_t)dX_t} \\ &= \frac{\int f(X_t)W(X_t)q(X_t|Z_{1:t})dX_t}{\int p(Z_{1:t}|X_t)p(X_t)\frac{q(X_t|Z_{1:t})}{q(X_t|Z_{1:t})}dX_t} \\ &= \frac{\int f(X_t)W(X_t)q(X_t|Z_{1:t})dX_t}{\int W(X_t)q(X_t|Z_{1:t})dX_t} \\ &= \frac{E_q[f(X_t)W(X_t)]}{E_q[W(X_t)]} \\ &\approx \frac{\frac{1}{n}\sum_{i=1}^{n} W(X_{t,i})f(X_{t,i})}{\frac{1}{n}\sum_{i=1}^{n} W(X_{t,i})} \\ &= \sum_{i=1}^{n} \bar{W}(X_{t,i})f(X_{t,i}) \end{aligned} \tag{3.63}$$

where $\bar{W}(X_{t,i})$ is the weight after the normalization

$$\bar{W}(X_{t,i}) = \frac{W(X_{t,i})}{\sum_{i=1}^{n} W(X_{t,i})} \tag{3.64}$$

The weight can be updated based on the following equation:

$$\begin{aligned} W_t(X_{t,i}) \propto \frac{p(X_{t,i}|Z_{1:t})}{q(X_{t,i}|Z_{1:t})} &= \frac{p(X_{t-1,i}|Z_{1:t-1})p(Z_{1:t}|X_{t,i})p(X_{t,i}|X_{t-1,i})}{q(X_{t,i}|X_{t-1,i}, Z_{1:t})q(X_{t-1,i}|Z_{1:t-1})} \\ &\propto W_{t-1}(X_{t-1,i})\frac{p(Z_{1:t}|X_{t,i})p(X_{t,i}|X_{t-1,i})}{q(X_{t,i}|X_{t-1,i}, Z_{1:t})} \end{aligned} \tag{3.65}$$

The particle filter suffers from a degeneracy problem. After a few steps, some particles with almost ignorable weights will cost calculation time although they do not provide much useful information for the position estimation. To improve the quality of the used particles, resampling is needed. The basic principle of resampling is to abandon the particles with ignorable weights and to copy the particles with large weights, as shown in Figure 3.5. Different methods can be used for resampling, such as residual resampling, multinomial resampling, stratified resampling, systematic resampling, rejection resampling and metropolis resampling. [PK].

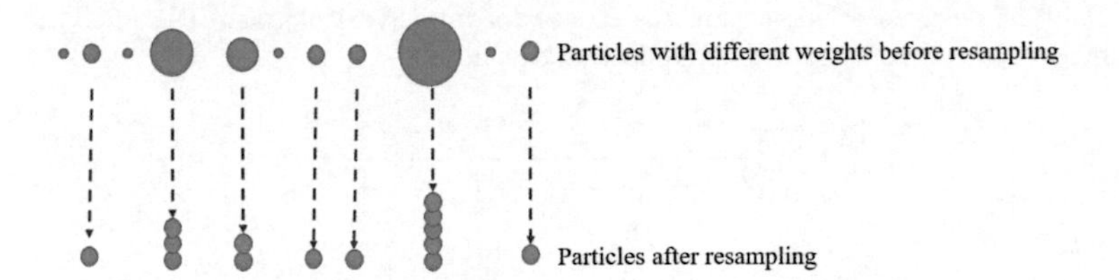

Figure 3.5: Resampling process

3.4.2 Particle Filter for Localization

Different to the KF, the particle filter can be used to solve non-linear and non-Gaussian noise problems based on the fact that the particles are used to describe the pdf. Compared to the KF, another large advantage of using the particle filter for NLOS detection based UWB localization is that the position estimation process is very robust with the frequent changing of the selected different BSs. The following introduces the particle filter process for TOA combined with NLOS detection. The particle filter with NLOS detection can be realized with the following steps: initialization; state transition, prediction; particle generation; NLOS detection; calculating weights for each particle; resampling; position estimation; and starting from state transition again. During the initialization step, the start position can be calculated based on the least squares method and the velocity can be set to zero. Similar to the KF, after initialization, the current state vector can be predicted based on the previous state vector. Also like the KF, the state vector X in the particle filter contains the position and velocity:

$$X_t = \begin{pmatrix} x_t \\ v_{xt} \\ y_t \\ v_{yt} \end{pmatrix} \tag{3.66}$$

$$u = [a_x \; a_y]' \tag{3.67}$$

where (x_t, y_t) is the position of the MS at time t, and (v_{xt}, v_{yt}) is the velocity of the MS at time t.

The following equation can be used to predict the current state:

$$X_t = AX_{t-1} + Bu + W \tag{3.68}$$

where W is the random errors for updating the particles. A is the state transition matrix, and B is called the control matrix.

Only the detected accurate ranges are used for further calculation. The predicted ranges can be calculated with the following equation:

$$D_{ij} = \sqrt{(x_{jt} - x_i)^2 + (y_{jt} - y_i)^2 + (z_{jt} - z_i)^2} \tag{3.69}$$

where the position of the ith BS is represented by $X_i = (x_i, y_i, z_i)$, while $X_s = (x_{jt}, y_{jt}, z_{jt})$ is the predicted position of the jth particle for the MS. The heights of the MS and BS are measured at the very beginning, which means that z_i and z_{jt} are constant values. D_{ij} represents the predicted ranges of the jth particle for the ith BS.

The weights (w_j) of the jth particle are calculated based on the error distributions and the differences (δ_{ij}) between the predicted ranges and the UWB measured ranges (D_{mi}).

$$\delta_{ij} = D_{mi} - D_{ij} \tag{3.70}$$

$$w_j = \prod_{1 \leq i \leq n} f(\delta_{ij}) \tag{3.71}$$

where n is the number of used BSs, and $f(\ldots)$ is the error distribution function.

During resampling, particles with low weights are replaced by those with high weights. Different methods can be used for resampling, such as systematic resampling, and residual resampling. After resampling, the particles are assumed to have the same weights. Thus the final estimated position and velocity are the average of the positions and velocities of these particles.

$$X_t = \frac{X_{t1} + X_{t2} + \cdots + X_{tN}}{N} \tag{3.72}$$

where N is the number of used particles and X_{tj} is the state vector of the jth particle at time t.

Finally, these particles are fed back to the state transition phase and used for the position estimation later on. The pseudocode of the particle filter algorithm is shown in Algorithm 1.

Algorithm 1 Particle Filter with NLOS detection Pseudo Code

1: Initialization
2: **for** t=2 $\rightarrow$ total time steps **do**
3:     **for** j=1 $\rightarrow$ used number of particles **do**
4:         $X_t = AX_{t-1} + Bu + W$
5:         Accurate ranges detection (D_{ij})
6:         $D_{ij} = \sqrt{(x_{jt} - x_i)^2 + (y_{jt} - y_i)^2 + (z_{jt} - z_i)^2}$
7:         $\delta_{ij} = D_{mi} - D_{ij}$
8:         $w_j = \prod_{1 \leq i \leq n} f(\delta_{ij})$
9:     **end for**
10:     Normalize weights
11:     Resample
12:     $X_t = \frac{X_{t1} + X_{t2} + \cdots + X_{tN}}{N}$
13: **end for**

3.5 Comparison of Filters

As previously discussed, the KF can be used to solve linear problems with Gaussian noise. With the Gaussian noise distribution propagation, the estimated error contains minimized variances. The EKF can solve non-linear problem with Gaussian noise. Based on the NLOS identification, the measurement noise ξ_i in R changes according to the identification results. During the real field testing, it is found that the EKF might suffer from a divergence problem with the frequent changes in the measurement noise. Furthermore, the stability of the EKF is not as good as that of the IEKF during the tests. This conclusion is purely based on real experiment data, collected with a UWB system in the Bosch Shanghai office. The particle filter can be used to solve non-linear problems with non-Gaussian noise based on the fact that almost all the noise distributions can be described with particles. Furthermore, during the tests, it is found that the particle filter is more stable compared to IEKF with the frequent changing of the used BSs. The main disadvantage of the particle filter is that the computation load is heavier than that of the IEKF. A comparison of these algorithms can be found in Table 3.1.

Table 3.1: Algorithm comparison

	Non-linear	**Computation load**	**Stability with NLOS detection**	**Non-Gaussian noise**
KF	Not suitable	+	Not suitable	+
EKF	+	++	+	+
IEFK	+	+++	++	+
PF	++	+++++	+++	+++

4 Hard NLOS Identification and Mitigation Approaches

As shown in the last chapter, the ranges under LOS are accurate. Even the measurements under glass, table and chair NLOS blockages are trustable. This kind of blockage is defined in this paper as soft NLOS (SNLOS). Based on the LOS and soft NLOS measurements, accurate position estimation can be achieved. However, in an indoor environment, the blockage can be human, water, metal or even a combination of these hard NLOS (HNLOS) blockages. Under these NLOS conditions, the measured ranges contain a non-ignorable positive bias that leads to inaccurate localization. Many algorithms have been developed to reduce the NLOS error, such as the Kalman filter and particle filter. However, even with the help of these algorithms, if inaccurate HNLOS measurements are used for position calculation, they still have great influence on the localization accuracy. The most effective method to reduce HNLOS error is to select the accurate measurements based on NLOS identification.

4.1 NLOS Identification

NLOS identification is used to select the HNLOS measurements. Several methods have been developed for this purpose (e.g. approaches based on range variance, maps, and CIR), as shown in Figure 4.1.

1) Range variance estimation based method: based on the fact that the variance for NLOS ranges is normally larger than that of LOS ranges, the running range variance is compared with a threshold to detect the NLOS measurements [SGKJ07], as shown

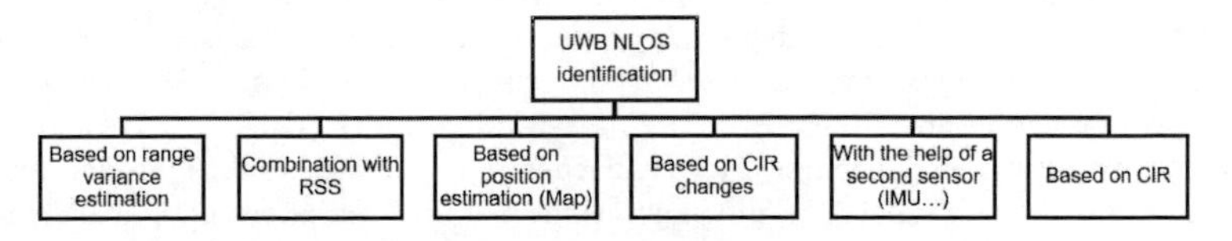

Figure 4.1: Overview of NLOS identification methods

in the following equation.

$$va = \frac{\sum_{n=1}^{N}(d_n - \mu_d)^2}{N-1} \text{[SGKJ07]} \tag{4.1}$$

where d_n is the range estimation, N is the used number of ranges, μ_d is the average of N ranges, and va is the defined running variance.

va is compared to the threshold. If the threshold is larger, then, it is LOS; otherwise, it is NLOS [SGKJ07]. The main drawback of the range estimation based detection method is the unavoidable additional latency due to the collection of the ranges. Furthermore, the identification accuracy is not very promising.

2) Combination of RSS: If a blockage exists between the MS and BS, the received signal strength (RSS) becomes smaller. NLOS detection can be realized based on the RSSs, as proposed in [XWM+15]. The DecaWave chips used in the developed system provide RSS information besides the range measurements. During the application of the RSS based method in this project, three main drawbacks are found. Firstly, the threshold used to compare with the RSS is very difficult to determine. With an improper threshold, the accuracy can be very poor. Secondly, the RSS is highly dependent on the indoor environments. With humans coming in or going out, the RSS could change dramatically. Furthermore, once the environment changes, the RSS changes totally. Thirdly, the NLOS identification accuracy based on this method is not very promising.

3) Map based method: Using a map, the NLOS measurements can theoretically be detected. The main disadvantage is that maps are not always available. Furthermore, except for fixed furniture, there are many moving objects or humans, and the NLOS detection accuracy based on the map is not very satisfying. Thus, this is not a good option and not the focus of this thesis.

4) CIR based NLOS identification: As discussed in the previous chapters, the CIRs for NLOS and LOS have different characteristics. Based on these differences, the NLOS CIRs can be detected using different algorithms, such as machine learning algorithms (SVM, artificial neural networks (ANN), etc.). The main advantage of this approach is the high NLOS detection accuracy. However, once the environment changes, to guarantee high accuracy, a new training dataset must be collected. This is highly time consuming. The machine learning CIR based detection contains three steps: feature extraction; model training based on the collected training dataset; and NLOS detection for new measurements based on the training model. This method has been intensively investigated in numerous papers [SH16], [Yan18], [GRS+17], [WYL17], [MGWW10]. Different features can be extracted from CIR, such as the kurtosis, crest factor, received signal energy. Although this method has been discussed in many papers, three important problems still need to be clarified. Firstly, different papers have used varying features, without explaining why these features are suitable for NLOS

detection. An overview of these features has not been given. Secondly, with different used features, the NLOS identification accuracy is also different. The best feature combination has not been discussed. Furthermore, for the machine learning algorithm, the influence of the parameters in machine learning for detection accuracy has not been discussed in detail. Thirdly, the accurate measurement selection differs between TOA and TDOA based on CIR. For TDOA, compared to the NLOS BS determination, the range difference selection based on two CIRs is more important. These problems are discussed into detail in the following chapters.

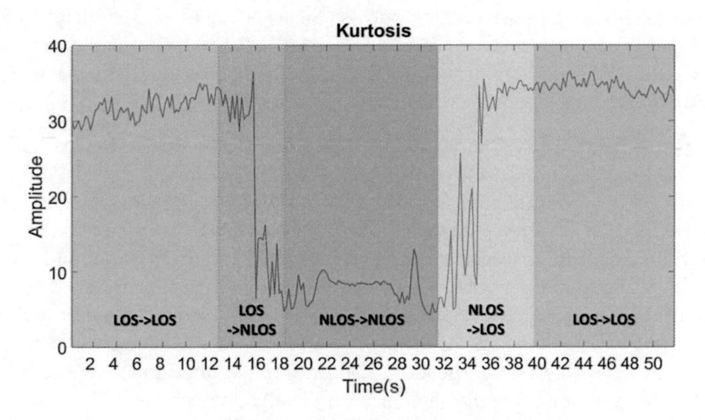

Figure 4.2: Kurtosis NLOS/LOS state change

5) CIR state change based method: The main drawbacks of the CIR based approaches are as follows. Firstly, huge amounts of training data are needed for accurate NLOS detection. Secondly, the training model cannot be universally used in different indoor environments. Once the indoor environment changes, a new training dataset needs to be collected to ensure accurate identification. Another possibility to realize NLOS identification is to detect the CIR state change. The CIR features during the state change from LOS to NLOS, or NLOS to LOS have significantly different characteristics compared to those during the change from LOS to LOS, or NLOS to NLOS, as shown in Figure 4.2 [ZYW+19]. Once the state change is identified, the NLOS measurements can be detected. [ZYW+19] proposes four different methods to detect the state change, based on the moving average (MA), slope, wavelet, and time series analysis. The main advantage of these methods are that they can be universally used in different environments for different UWB systems. Furthermore, no huge dataset is needed for training. However, determining the proper threshold is highly challenging. The main problem is that once wrong detection happens, the following state detection will be all false until the second wrong detection occurs. Thus, this method can be used as a second method to improve NLOS identification accuracy together with an-

other NLOS detection method, but it is not suggested to detect NLOS measurements solely based on this approach.

6) IMU based NLOS identification: The combination with a second sensor source, like an inertial measurement unit (IMU), is a widely used method for UWB NLOS identification and mitigation [BN18], [XTC18]. A tightly coupled UWB/IMU approach is proposed in [HDLS09], which utilizes the prediction of the IMU measurements to detect the inaccurate range measurements. Since the range error has a non-Gaussian distribution under NLOS, one of the drawbacks of this approach is the violation of the precondition for the proposed error model, so the accuracy of the identification for the NLOS range measurements cannot be guaranteed. Chapter 7 presents a new method for accurate UWB measurements selection with the help of the IMU measurements based on the triangle inequality theorem.

A comparison of these NLOS identification methods is presented in Figure 4.3. Based on the NLOS detection accuracy and the engineering feasibility, the CIR based- and IMU based NLOS identification are two of the best approaches. Compared to the IMU based method, the CIR based method is less complex. Once the proper training model is built, the localization accuracy can be dramatically improved based on NLOS detection with the training model. However, the computation load is very high, since large CIR datasets need to be processed. The IMU based method, in contrast, does not need to process huge amounts of data, but the NLOS identification accuracy is high depending on the algorithms. Most of the time, a proper threshold needs to be determined. How to optimize the threshold is a very challenging problem. The localization accuracy improvement with IMU/UWB fusion highly depends on the developed algorithms.

	Approach	Algorithm	Assumptions	Accuracy	Engineering Feasibility	Ranking
1	Based on range estimation	Running Variance	Variance of the Ranges change dramatically under NLOS	Low	Low/Medium Large latency	3
2	Based on CIR	1.ANN, SVM... 2.Probability density function	CIRs are stable and can be clearly identified into LOS/NLOS conditions	High	High Huge computation load;	1
3	Based on position estimation (Map)	With the help of the map	Environment does not change dramatically	Low/Medium	Medium	3
4	Based on CIR changes	1. Time series analysis 2. CIR change detection (ANN, SVM...) 3. Static LOS/NLOS collection and comparison	The CIR change from LOS to NLOS and the other way around are obvious	Medium (Can only be used as reference)	Medium	3
5	With the help of a second sensor (IMU...)	EKF, PF etc.		Medium/High	High	1
6	Combination with RSS	Received signal strength	Signal strength changes dramatically under NLOS	Low	Medium	3

Figure 4.3: Comparison of the NLOS identification algorithms

4.2 NLOS Mitigation

If NLOS measurements can be identified, NLOS mitigation can be achieved by solely using the detected accurate measurements for position estimation. To further improve the localization accuracy, different optimization algorithms can be used. The Kalman filter and particle filter describe the dynamic process of the position estimation and can easily be combined with the NLOS identification method. Thus, they are highly suitable to solve the UWB localization problem. These filter algorithms were discussed in Chapter 3.

5 UWB Localization in Office Environment

The indoor localization performance is influenced by the environment. In this chapter, the UWB localization performance is evaluated in the Bosch Shanghai office environment. The office map is presented in Figure 5.1. The data for analysis is collected during the night when no workers are present. The localization is done during the day with people walking around. Two different situations are considered here: first, the localization is realized with three different BSs, and the localization accuracy is improved with the noise model; and second, eight BSs are used, and the localization accuracy is improved based on NLOS identification.

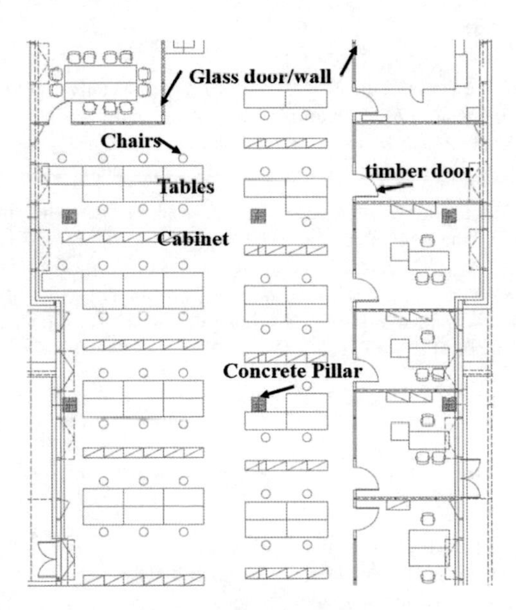

Figure 5.1: Office environment

5.1 Localization with Three BSs

During the experiments, the heights of the BSs and MSs are fixed so that the localization can be simplified into a 2D case. At least three BSs are needed for position estimation. Without any redundant BSs, the NLOS identification is not very useful. Based on the proper noise model, the localization performance can be improved. As discussed before, due to the dramatic change in noise in different environments, a general noise model can not be built. Furthermore, the LOS or NLOS noise distribution at a fixed position should not be used to describe the noise for the whole environment, since in different positions the noise distributions are different. Noise distributions even differ for different MSs/BSs under the same conditions. During the experiments, the error distributions for three BSs are built. The MS is randomly fixed in more than 80 LOS and 50 NLOS positions. The reference ranges are measured by the Bosch GLM 100 C Professional laser measure. In a dynamic process, the MS is held by a human and moved randomly in the office. The reference positions are measured by cameras, so that the reference ranges can be calculated. The error distribution is built based on these reference measurements and UWB measurements. A stable distribution is found to be highly suitable to describe the error distribution.

Stable Distribution

Due to the NLOS error, the error distribution has heavy tails. Furthermore, since the NLOS errors contain positive biases, the right tail is longer, which can be called right-skewed. A stable distribution is very suitable to model heavy tails and skewness distribution. The definition of a stable distribution is presented below.

Assuming that $X_1, X_2, ..., X_n$ are independent variables and have the same distribution as X, if the sum of these independent variables $(X_1 + X_2 + ... + X_n)$ has the same distribution as $b_n X + c_n$, where $b_n > 0$ and $c_n \in (-\infty, +\infty)$. Then, the distribution of the random variable X is a stable distribution. The probability density function of a stable distribution is defined by its Fourier transform, which is called the characteristic function $\varphi(t)$ [LLZP07],[PSWW12].

$$\varphi(t) = \begin{cases} exp(-\gamma^{\alpha}|t|^{\alpha}(1 - i\beta tan\frac{\pi\alpha}{2}sign(t)) + it\delta), \alpha \neq 1; \\ exp(-\gamma|t|(1 + i\beta\frac{2}{\pi}sign(t)log|t|) + it\delta), \alpha = 1. \end{cases} \tag{5.1}$$

Where
(1) $\alpha \in (0, 2]$ describes the tails of the distribution;
(2) $\beta \in [-1, 1]$ represents the skewness of the distribution;
(3) $\gamma \in (0, +\infty)$ is the scale parameter;
(4) $\delta \in (-\infty, +\infty)$ is used to shift the distribution to the left or right. The probability

density function can be calculated with the following equation:

$$f(x;\alpha,\beta,\gamma,\delta) = \frac{1}{2\pi}\int_{-\infty}^{+\infty} \varphi(t)e^{-itx}dt \quad (5.2)$$

The probability density function of a stable distribution with different β is shown in Fig. 5.2.

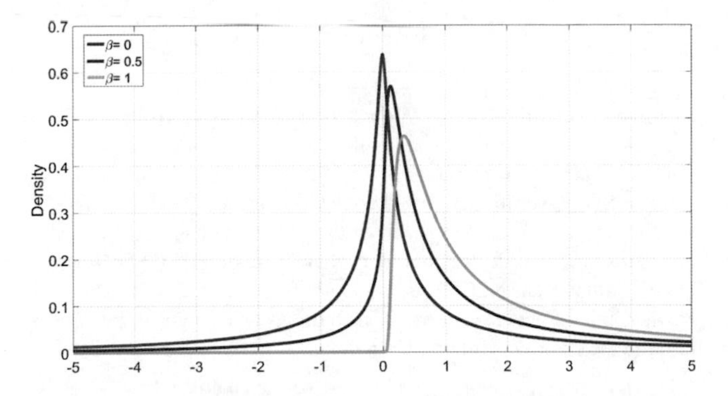

Figure 5.2: Probability density function of a stable distribution with different β

After combining the LOS/NLOS errors, the office error distribution is built. Figure 5.3 shows the error distributions for the first and the second BS. These distributions can be modeled as stable distributions. It can be observed that for different BSs, the error distribution varies. Thus, a separate distribution model for different BSs is necessary. Since the error distribution is non-Gaussian, the particle filter is more suitable to solve this problem. The pseudocode of the particle filter algorithm was already shown in Algorithm 1. The weights (w_j) of the jth particle are calculated based on the office error distributions:

$$w_j = \prod_{1\leq i\leq 3} f(\delta_{ij};\alpha,\beta,\gamma,\delta) \quad (5.3)$$

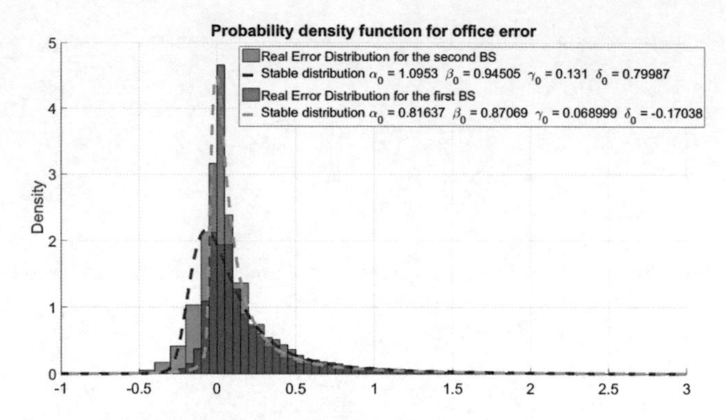

Figure 5.3: Office error distributions model

The BSs are fixed in the following positions: BS1=[0 0], BS2=[0 18.34], BS3=[-11.45 18.34]. Fig. 5.4 shows the localization results with the Taylor series method, the particle filter with normal error distribution (PF ND), and the particle filter with stable error distribution (PF SD). Overall, the PF SD (black points) has better accuracy especially in the red dotted square areas. However, in the blue dotted square areas, the position estimation results based on PF ND are better than those of the PF SD. The main reason for this is that in that area, NLOS barely occurs for BS1 and BS2, which means that the normal distribution is the real error distribution. Thus, a general error distribution cannot describe the error characteristics in detail.

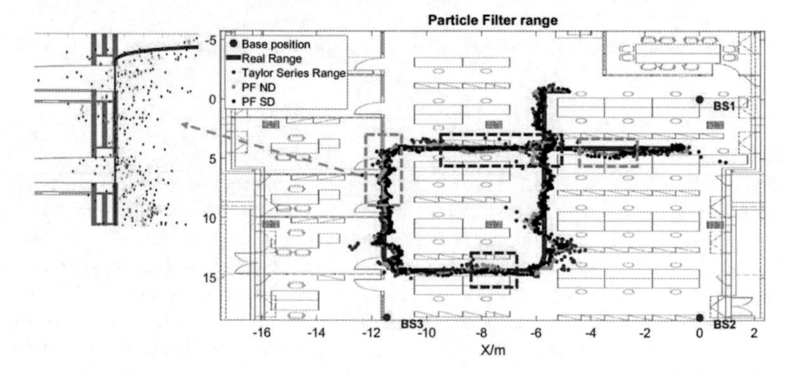

Figure 5.4: Localization results with Taylor series method, PF ND and PF SD

To obtain more accurate error distribution models, the whole floor is divided into four

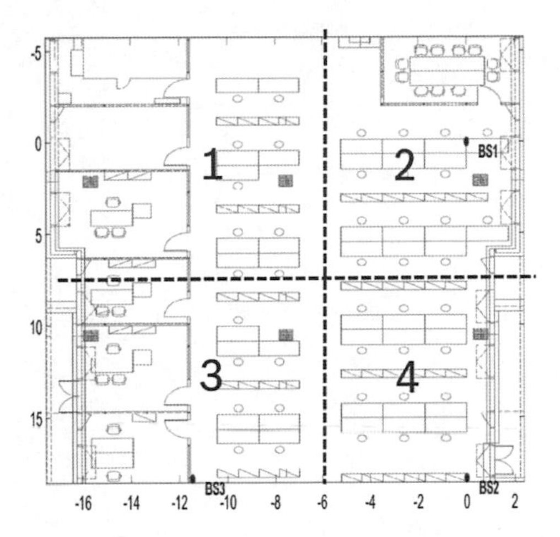

Figure 5.5: Office map divided into four different areas

different areas (as shown in Figure 5.5) based on the NLOS frequency. For example, in area 2, NLOS never happens for BS1, whereas, BS3 is frequently under NLOS due to the blockage by the cylindrical wall and the thin metal cabinets. Thus, the normal distribution model is built for BS1 and a stable error distribution model is built for BS3 in area 2. For each area, a separate error model is built for each BS. During the position estimation with the particle filter, the previous position determines in which area the MS is, and which error distributions need to be used. The localization results based on the particle filter with different error distribution models (ESD based PF) are compared with the PF ND and PF SD in Figure 5.6. Overall, the ESD based PF shows better position estimation results, especially in the red dotted square areas. This proves that the stable distribution error model can be used in the particle filter to improve the UWB localization accuracy. However, while the errors are smaller with these models, the calculated positions still have small deviations from the real positions, such as the red dotted square areas in Figure 5.6.

5.2 Localization with Redundant BSs

Although localization performance can be improved with proper error distribution models, the NLOS measurements still influence the position estimation accuracy. A better solution to further improve accuracy is to add redundant BSs and to use only the selected accurate ranges based on NLOS identification for further calculation. As

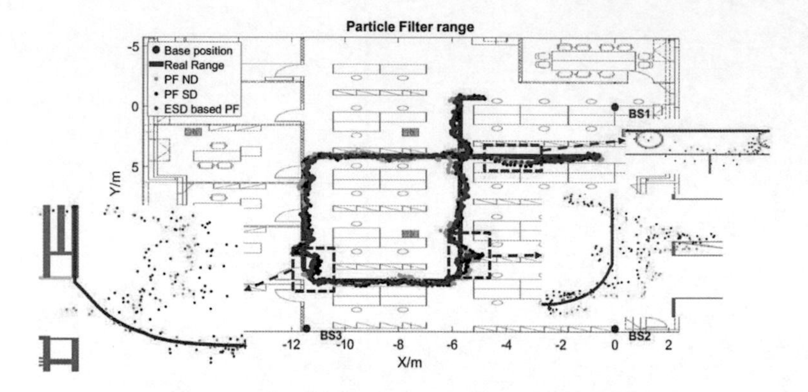

Figure 5.6: Localization results with PF ND, PF SD and ESD based PF

previously mentioned, CIRs differ based on the environment. The testing environment is the Bosch Shanghai office. The typical CIRs in the office environment were already presented in Figure 2.27. Different features can be extracted based on the differences in these CIRs. Inaccurate measurements are caused by the NLOS error in the testing environment. Thus, if the NLOS CIRs can be detected, the inaccurate ranges can be selected out.

5.2.1 Extracted Features from CIRs

Five different feature groups are identified based on the distance, CIR shape, time, multipath richness and power. In each group, different features that depend on the same factor are determined. One of the features from each group is selected for machine learning.

Distance Based Factors

At the same distance and in the same environment, the energy and maximal amplitude of the CIRs for accurate ranges are larger than those of the NLOS CIRs. These features can be calculated using the following equations:

$$E = \int_{-\infty}^{+\infty} |c(t)|^2 dt \tag{5.4}$$

$$c_{max} = max\ c(t) \tag{5.5}$$

where E is the received signal energy and c_{max} is the maximal amplitude. However, as the distance between the MS and BS becomes larger, the energy and maximal amplitude of the CIR decrease. Thus, these two features must be used along with the distance for machine learning. Due to the multipath effect, the maximal amplitude of the CIR under NLOS could be even larger than that of the CIR for an accurate range. At the same time, based on the experimental results, it is unlikely that the energy under NLOS will be larger than under LOS. Hence, it is better to use energy as a feature for machine learning. However, under severe multipath, the energy of the NLOS CIR can still be greater than that of the CIR for accurate ranges with longer distances, as shown in Figure 5.7. The shapes of these CIRs (Figure 5.7) are different. These CIRs can be differentiated using shape based factors.

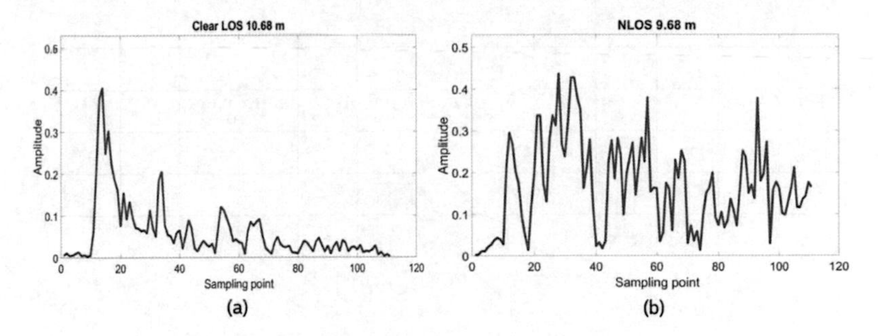

Figure 5.7: CIR (**a**) under clear LOS at a distance of 10.68 m; (**b**) under NLOS and multipath at a distance of 9.68 m.

Shape Based Factors

Five features can be used to describe the shape of the CIRs: standard deviation, kurtosis, crest factor, form factor and skewness. The standard deviation describes how close the data points are to the mean of these data. The kurtosis represents the "tailedness" of the curve. The crest factor and the form factor indicate the extremeness of the peaks in the curve. The skewness can be used to differentiate the shape of the curve based on the distribution of the tail. These features can be computed with the following equations:

1) Standard deviation: σ

2) Kurtosis

$$\kappa = \frac{E[(|c(t)| - \mu)^4]}{\sigma^4} \tag{5.6}$$

where μ denotes the mean of c(t).

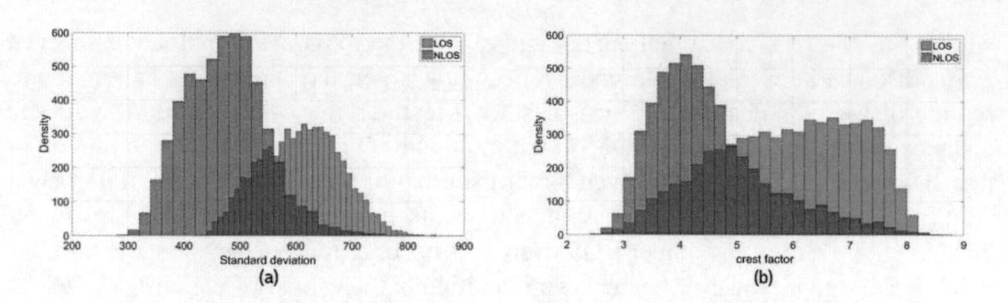

Figure 5.8: Histogram for (**a**) standard deviation; (**b**) crest factor.

3) Form factor

$$k_f = \frac{C_{RMS}}{\overline{C}} \tag{5.7}$$

where C_{RMS} is the root mean square of the signal and $\overline{C}$ is the average of the signal.

4) Crest factor

$$C_f = \frac{c_{max}}{C_{RMS}} \tag{5.8}$$

5) Skewness

$$\gamma = \frac{E[(|c(t)| - \mu)^3]}{\sigma^3} \tag{5.9}$$

The histogram of the CIRs for accurate ranges and NLOS CIRs can be used to examine which features might be more effective to differentiate. Figure 5.8 shows the histogram for standard deviation and crest factor. The smaller the intersection area of the CIRs for accurate ranges and the NLOS CIRs is, the better the feature is. Compared to the other shape based factors, the standard deviation shows the best results. However, the shapes of the CIRs for accurate ranges and the NLOS CIRs can be similar, if the pulse from the direct path is totally blocked and the reflected signals are mainly received in the first half of the CIR, as shown in Figure 5.9. The shapes of these CIRs are not exactly the same, but these shape based factors for these CIRs are not so different. Other possible features to differentiate these features are time based factors.

Time Based Factors

Two time based features can be used for NLOS CIR detection: rise time to the maximal amplitude, and peak to start of the received pulses time delay (PFTD). Due to the blockage in the direct path, the first arrived path is not the strongest one. Instead, the strongest path happens later due to the multipath effect. However, with ignorable blockages, such as thin glass, although the first path is not the strongest path, the

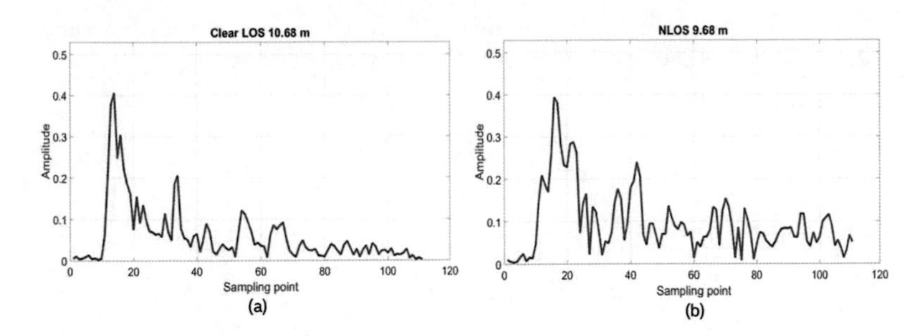

Figure 5.9: CIR (**a**) under clear LOS at a distance of 10.68 m (**b**) under NLOS at a distance of 9.68 m.

measured ranges are still accurate, as shown in Figure 5.10.

$$PFTD = \tau_{strongest} - T_{sn} \tag{5.10}$$

where $\tau_{strongest}$ is the arrival time of the strongest path and T_{sn} is the start time of the arrived signal. A threshold is predefined to determine those paths. If the arrival signal amplitude is larger than this threshold, the real arrived signal starts.

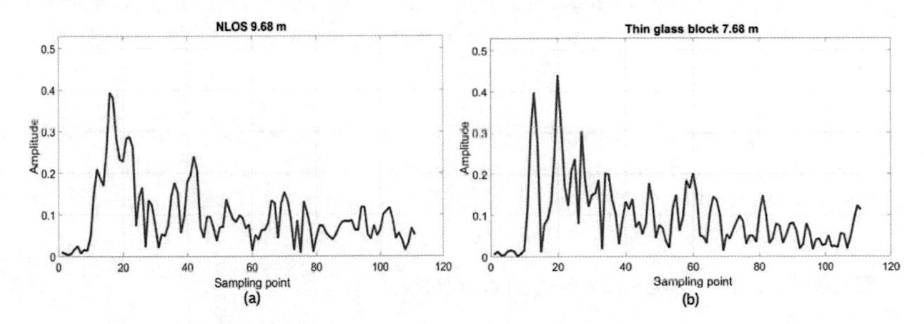

Figure 5.10: CIR (**a**) under NLOS at a distance of 9.68 m; (**b**) under thin glass blockage at a distance of 7.68 m.

Multipath Richness Based Factors

The mean excess delay and the root-mean-square (RMS) delay spread indicate the multipath richness. These two features provide additional information about the CIRs. Since NLOS occurs more frequently in rich multipath areas and in different rich multipath areas, the multipath part of the CIRs is different. Thus, the use of these features can help to improve the NLOS detection accuracy. The mean excess delay and the RMS delay spread can be calculated with the following equations:

1) Mean excess delay

$$\tau_m = \frac{\int_{-\infty}^{+\infty} t|c(t)|^2 dt}{E} \tag{5.11}$$

2) Root-mean-square (RMS) delay spread

$$\tau_{RMS} = \sqrt{\frac{\int_{-\infty}^{+\infty} (t-\tau_m)^2 |c(t)|^2 dt}{E}} \tag{5.12}$$

Power Based Factors

There are many power related factors that can be used for NLOS detection such as the power difference between the first path and the strongest path; the power ratio of the first path and the strongest path; the power difference between the total power and the first path power; the power ratio of the total power and the first path power; the signal to noise ratio (SNR); the peak to average power ratio. However, these factors are correlated with the features mentioned above, which means that the changing tendency is similar to that of the other features. Thus, it is not necessary to combine power based factors with other features.

Based on the theoretical analysis, the best feature combination for the NLOS identification in the Bosch Shanghai office may be distance, energy, standard deviation, peak to start of the received pulses time delay and the mean excess delay/RMS delay spread.

5.2.2 Effective Signal Length of CIRs

During the experiments, the CIRs are recorded as discrete signal sampling points, not as analog signals. For the developed UWB system, one CIR sample received by the BS with the Decawave chip provides 1,015 discrete points. Only a small portion of these points (marked in red) represents the real received impulse signal, while the rest (marked in blue) are just noise, as shown in Fig. 5.11. The CIRs presented in the above chapters are only the real received impulse signal. However, even not all the sampling points for the real received impulse are needed, since only part of the received impulse signals are different, while the rest are quite similar. In Fig. 5.12, the received impulse signals under LOS and NLOS are presented. As shown in this figure, the most significant difference between LOS and NLOS CIRs is observed in the first half of the

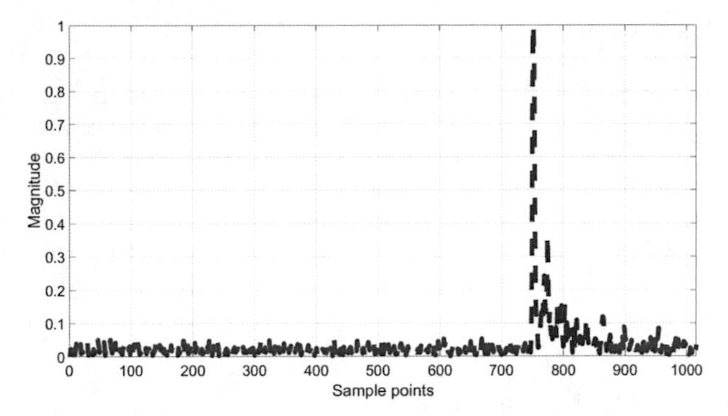

Figure 5.11: Original CIR sample with 1,015 points

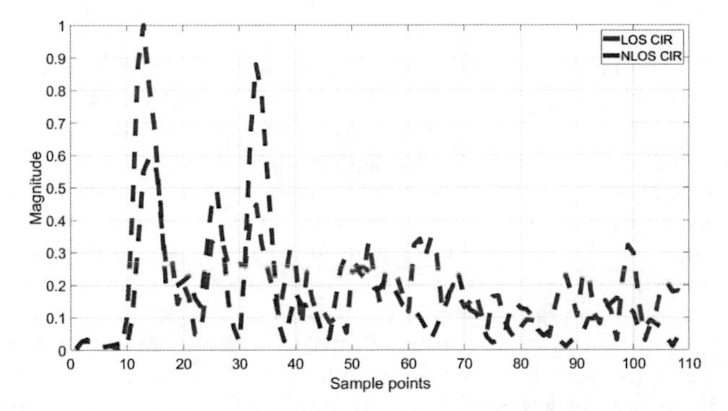

Figure 5.12: CIRs under LOS and NLOS

real received impulse signals, while the rest of these points are similar. The calculation complexity and system delay in a real time application can be reduced by using fewer CIR points. Thus, only those points in the real received impulse signal that reflect the most significant difference are selected and used for classification. These points are defined as the effective signal length of the CIR. During the system configuration, once the effective signal length can be predefined, only these CIR points are saved and used for calculation. The fewer CIR points are used, the lower is the calculation complexity.

5.2.3 Test Results

During the field tests, the performance of the NLOS identification accuracy based on SVM with different parameters is evaluated. The NLOS detection accuracy is compared using different feature combinations after tuning the parameters. After the feature combination is optimized, the effective signal length of the CIRs is determined. The position estimation is realized with eight BSs. Compared to the standard particle filter and the Taylor series based method, the particle filter with NLOS detection has better accuracy, since the inaccurate ranges are selected out and not used for position estimation.

SVM Classification Performance

To determine the feature combination for NLOS detection, the data are collected with one UWB BS and one MS. The Decawave DW1000 is the chip used in the UWB system. The training dataset, which contains the measured range, the CIR and the label indicating whether the CIR is measured under LOS, is recorded in different places with different distances between the BS and the MS in the office environment. The Bosch GLM 100 C Professional laser measure is used to measure the reference ranges. The training data are divided into two groups: the CIRs for accurate ranges and NLOS CIRs. During the data collection phase, it is found that sometimes even under clear LOS, the range measurements are inaccurate and the corresponding CIRs look very similar to the NLOS CIRs. One possible explanation for this is that the random errors occur when the signal is transmitted from the MS to the BS. These CIRs are selected out and are not used in the training dataset. To build the correct training dataset, the following conditions are used to separate these two groups: The ranges which belong to the CIRs for accurate range group: firstly, the difference between the reference range and the UWB measured range is smaller than 30 cm; secondly, the range is measured under clear LOS, with a thin glass, paper or table etc. blockage. As for the CIRs in the NLOS groups, the ranges have to be measured with blockage, and the difference between the reference range and the UWB measured range is larger than 30 cm. After the SVM model is trained, the inaccurate ranges can be detected.

More than 20,000 CIRs are recorded to train the SVM model, and another 10,000 CIRs are used to evaluate the trained model. Two parameters (P_{LOS} and P_{NLOS}) are used to evaluate the identification results:

$$P_{LOS} = \frac{CL}{AL} \tag{5.13}$$

$$P_{NLOS} = \frac{CN}{AN} \tag{5.14}$$

where CL is the number of correctly detected LOS CIRs, and AL is the total number

Table 5.1: P_{LOS} for different C and Gamma with 110 CIR points

C/Gamma	0.1	0.2	1	2	10	20	100	200
0.002	49.64%	61.6%	92.16%	92.74%	95.07%	96.62%	97.01%	97.01%
0.01	49.64%	65.1%	93.89%	94.75%	93.52 %	96.28%	97.01%	97.01%
0.02	49.64%	76.83%	95.7%	95.89%	94.31%	93.5%	97.01%	97.01%
0.1	65.7%	90.15%	96.9%	96.84%	95.79%	94.48%	97.01%	97.01%
0.2	77.18%	93.93%	96.9%	96.87%	96.44%	95.08%	96.77%	97.01%

Table 5.2: P_{NLOS} for different C and Gamma with 110 CIR points

C/Gamma	0.1	0.2	1	2	10	20	100	200
0.002	99.35%	98.84%	91.83%	87.77%	61.36%	51.77%	48.11%	47.77%
0.01	99.35%	98.46%	93.91%	92.7%	78.67%	52.68%	48.11%	47.77%
0.02	99.35%	97.82%	94.31%	93.58%	83.41%	71.11%	48.11%	47.77%
0.1	98.41%	96.14%	94.83%	94.6%	92.26%	85.45%	48.11%	47.77%
0.2	97.84%	95.83%	94.89%	94.78%	93.38%	89.49%	48.75%	47.77%

of the testing LOS CIRs. CN is the number of correctly detected NLOS CIRs and AN is the total number of the testing NLOS CIRs.

Two parameters need to be tuned to improve the classification accuracy: C and Gamma. Table 5.1 and 5.2 show the value of P_{LOS}/ P_{NLOS} with 110 CIR points for different Cs and Gammas with the distance, energy, standard deviation, PFTD and RMS delay spread as the feature combination. The first column shows the value of different Cs and the first row indicates the values for different Gammas. By increasing Gamma, P_{LOS} becomes larger but P_{NLOS} becomes smaller. The overall accuracy that achieves the best performance is in the last row and the fourth column. The P_{LOS} and P_{NLOS} are equal to 96.9% and 94.89%.

The identification accuracy of different feature combinations is also checked. As previously mentioned, these features are extracted based on the following related factors. Distance based factors: 0. distance between BS and MS, 1. energy, 2. maximal amplitude; shape based factors: 3. standard deviation, 4. kurtosis, 5. form factor, 6. crest factor, 7. skewness; time based factors: 8. rise time to the maximal amplitude, 9. PFTD; multipath richness based factors: 10. mean excess delay, 11. RMS delay. Table 5.3 shows the accuracy of different feature combinations after parameter optimization. 0,1,3,9,10 means that the feature combination contains distance between BS and MS, energy, standard deviation, PFTD and mean excess delay. The best accuracy for P_{LOS} is the FC in the second column, while for P_{NLOS} it is in the fifth column. As shown in Table 5.3, the accuracy of the feature combination with at least one feature used in each related factors group is better than the accuracy of other combinations.

Table 5.3: P_{LOS} and P_{NLOS} with 110 CIR points for different feature combinations (FC) after the parameter tuning

P/FC	0,1,3,9,11	0,1,3,9,10	0,1,3,4,9,10	0,1,3,4,9,10,11	0,1,3,9	0,1,3,10	0,1,...,10,11
P_{LOS}	96.9%	94.04%	95.14 %	96.12%	91.05 %	90.33%	96.53%
P_{NLOS}	94.89%	91.21%	92.37%	94.93 %	90.95%	91.15%	94.73%

Table 5.4: P_{LOS} and P_{NLOS} with different CIR length after the parameter tuning

P/CIR length	1 to 40	1 to 60	1 to 80	7 to 60	7 to 80
P_{LOS}	68.19%	89.58%	97.11%	95.45%	96.61%
P_{NLOS}	76.84%	88.51%	95.13%	92.95%	92.53%

The main reason for this is that the features in each related factors group can detect the NLOS CIR under at least one specific situation. With more than one feature used in each group, the accuracy does not become much better; it stays the same or even decreases. The best overall accuracy is achieved with the feature combination using the distance, energy, standard deviation, PFTD and RMS delay spread as a feature combination (0,1,3,9,10).

After the best feature combination is determined, the influence of the length of the used CIR points for the NLOS identification is evaluated. The CIRs with different lengths of the sampling points are investigated: starting from 1 to 110, 100 ... 40; 4 to 110, 100 ... 40; ... 19 to 110, 100 ... 40. Table 5.4 shows the detection accuracy with different CIR lengths after the optimization of the SVM parameters. The optimal effective signal length of the CIR ranges from 1 to 80. The P_{LOS} and P_{NLOS} are equal to 97.11% and 95.13%.

Localization Performance

A total of eight BSs are used to evaluate the position estimation algorithm based on the particle filter with NLOS identification. Only the detected accurate ranges based on the SVM model are used for further calculation. These BSs are preinstalled in the following positions: BS1=[0 0], BS2=[0 8.31], BS3=[-6 13.03], BS4=[0 18.34], BS5=[-6 6.44], BS6=[-11.45 18.34], BS7=[-11.45 8.34], BS8=[-11.45 0]. Figure 5.13 shows the localization results with the standard particle filter, the Taylor series method and the particle filter with NLOS identification. Compared to the other two methods, the particle filter with NLOS detection (blue points) has better accuracy, as shown in the figure. In the right corner of the walking trajectory, BS 8 and BS 7 are blocked by the cylinder wall. BS 6 and BS 1 are blocked by humans. This is the reason for the inaccurate position estimation with the Taylor series method and the particle filter. With the SVM model, the accurate ranges can be selected and only these accurate measurements are used for calculation. Thus, the localization accuracy of the particle filter with NLOS detection can maximally mitigate the position error caused by NLOS bias.

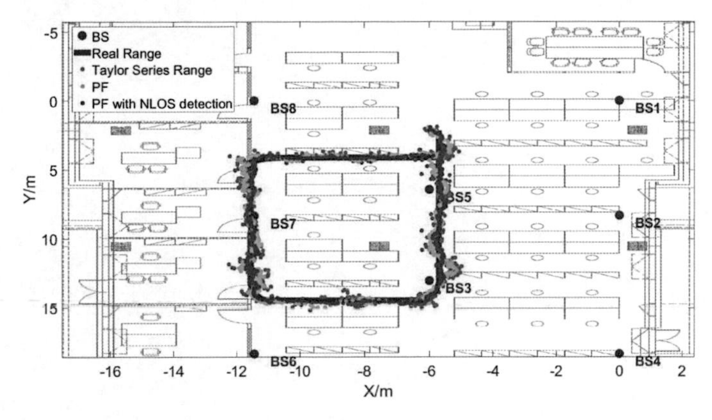

Figure 5.13: CIR (**a**) under NLOS at a distance of 9.68 m; (**b**) under a thin glass blockage at a distance of 7.68 m.

5.3 Summary

The UWB localization in the office environment stays in the focus in this chapter. The localization accuracy with three BSs can be improved with a proper error distribution model (stable distribution model in the Bosch Shanghai office environment). However, the NLOS condition still cause inaccurate position estimation. A solution to further improve the accuracy is to add redundant BSs and to use only the selected accurate ranges based on NLOS identification for further calculation. The SVM algorithm is used for NLOS detection. The feature combination, the used CIR length and the parameters in SVM are optimized to improve the identification accuracy. The position estimation accuracy is improved with the SVM based NLOS detection approach in the Bosch Shanghai office environment.

6 UWB Localization in Harsh Industrial Environment

One of the main conditions under which NLOS detection can be used to improve localization accuracy in the TOA approach is when at least three BSs are detected as accurate. However, in harsh industrial environments, it happens frequently that less than two ranges are measured under LOS. Thus, the NLOS identification based TOA approach might not work well. To check the performance of this approach, the position estimation accuracy is evaluated in the Bosch Changsha plant. The map of the industrial floor is presented in Fig. 6.1. A typical machine environment is shown in Fig. 6.2. Seven BSs are installed in the predefined positions : BS1=[0.3 -0.3], BS2=[13 -0.3], BS3=[13, -4.3], BS4=[13 -8.6], BS5=[0.3 -8.6], BS6=[0.3 -4.3] and BS7=[6.8 -2.3]. In most positions, a maximum of two BSs are under clear LOS, while the remaining BSs receive the signal with plastic combined with metal and human blockages. It can even happen that none of the BSs is under LOS. The same algorithms are used as in the previous chapter.

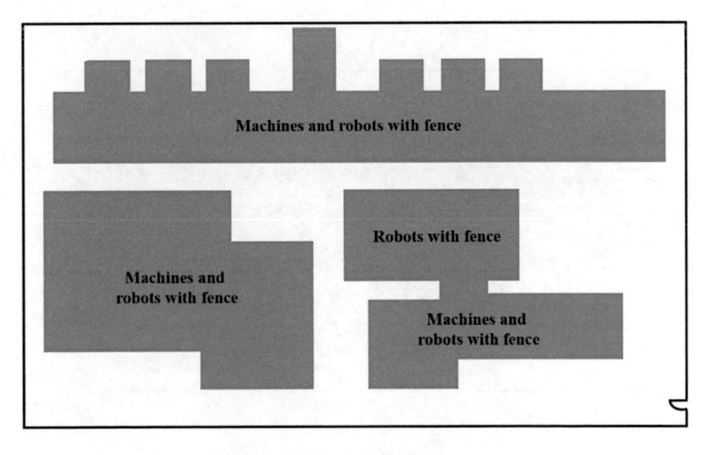

Figure 6.1: Industrial floor map

Figure 6.2: Typical real industrial environment

Fig. 6.3 shows the localization results using the Taylor series, the normal TOA based on the particle filter and the TOA based on the particle filter with NLOS range detection. The red line is the predefined trajectory, while the black points are the localization results with the TOA based on the particle filter with NLOS range detection. As shown in this figure, the position estimation accuracy with the NLOS range detection is slightly better than the other two approaches. However, the NLOS error can not be reduced in many places, especially around the upper-right corner. The main reason for this is that during the test, usually a maximum of two or even no BSs are identified as LOS.

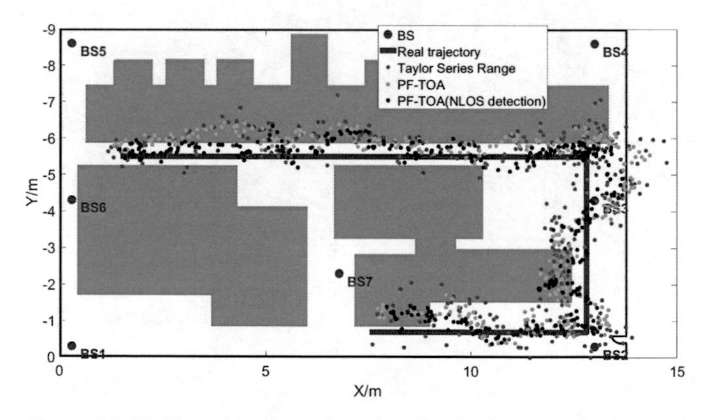

Figure 6.3: Field test localization results with the TOA based approaches

Compared to TOA, the range difference based TDOA approach might be more suitable for harsh industrial environments. In theory, the range difference can be directly calculated with two different ranges. If both ranges are accurate, then the calculated range difference is also accurate. If only one of the BSs is obstructed, all the range differences calculated with this BS and an accurate range contain a non-negligible positive bias. However, if both ranges are measured under NLOS and it coincidentally happens that the same delay is contained in these measurements, then the range difference is again accurate due to the compensation of the errors. Two NLOS ranges lead to an inaccurate range difference, only if the NLOS biases are different. In a 2D case, accurate localization can be realized with two range differences. The difference between TOA and TDOA can be further explained as follows. As discussed earlier, the range (D_i) between the ith BS and MS in the ideal case is:

$$D_i^2 = \sqrt{(x_i - x_s)^2 + (y_i - y_s)^2 + (z_i - z_s)^2} \tag{6.1}$$

where the position of the ith BS is represented by $X_i = (x_i, y_i, z_i)$ and $X_s = (x_s, y_s, z_s)$ is the real position of the MS. z_i and z_s are constant, since the height of the MS and BS are fixed. Thus, the position estimation can be simplified into a 2D problem.

With at least three different range measurements, the position of the MS can be calculated. In a 2D case, the position of the MS must lie on the circle that is centered at the BS. In the ideal case, the MS position is the unique intersection point of at least three different circles.

In real cases, the range measurements (D_{mi}) contain system noise (ε_i). Furthermore, if the BS is under NLOS, a bias (b_i) is added to the range. The bias has a huge influence on the localization accuracy.

$$D_{mi} = D_i + \varepsilon_i + b_i \tag{6.2}$$

At least three accurate ranges are needed for accurate position estimation for TOA. In harsh industrial areas, it can happen that in most positions a maximum of two or even no BSs are under LOS. Accurate position estimation with TOA is not possible under such conditions.

In contrast to TOA, the range difference is used for position estimation in TDOA. The TDOA position estimation principle can be described as follows. First, range differences need to be obtained. Then, the intersection of the hyperbolas, that are generated based on the range differences with foci at the BSs, is the position of the MS. If the signal propagation is under LOS, the accurate range difference is:

$$\Delta d_{ij} = D_i - D_j + \varepsilon_i - \varepsilon_j \tag{6.3}$$

Based on the Taylor series method, $D_i - D_j$ can be expanded into the Taylor series. After the expansion, the position of the MS can be calculated based on the LS method with at least two range difference measurements.

If the signal propagation is under NLOS, the range difference is:

$$\Delta d_{ij} = D_i - D_j + \varepsilon_i + b_i - \varepsilon_j - b_j \tag{6.4}$$

where b_i is the bias caused by NLOS for the ith BS and b_j is the bias caused by NLOS for the jth BS.

If these two errors happen to be almost the same, then $\varepsilon_i - \varepsilon_j$ and $\varepsilon_i + b_i - \varepsilon_j - b_j$ can be in the same range due to the compensation of these errors. Thus, the range differences calculated with the ranges under LOS or NLOS conditions can be accurate and can be used for accurate position estimation. This means that if the identification of the accurate range difference (calculated with two NLOS ranges) is possible, accurate localization can be realized even with the NLOS ranges.

6.1 Overview of the TOA/TDOA Combination Approach

A TOA/TDOA combination approach is proposed and used to improve the localization accuracy in harsh industrial environments in this Chapter. This approach can be achieved in three steps:

1. Off-line SVM model training

Firstly, different SVM models are trained based on the collected dataset. The first kind of SVM model is used to select the accurate range measurements, while by utilizing the compensation of the bias errors, the accurate range differences are selected with the second kind of SVM model.

2. On-line accurate measurement selection

If at least three ranges can be identified as accurate with the first kind of trained models, these LOS ranges are directly used for position estimation. Otherwise, the range differences are calculated with the detected inaccurate ranges, and the second kind of SVM model is activated to select the accurate range differences.

3. On-line position estimation

Since the used BSs and the measurements changes frequently, the particle filter is more stable than the KF, as previously mentioned. Furthermore, the accuracy of the accurate range difference detection is below 85%. With the consideration of the inaccurate range differences that are identified as accurate, the error distribution cannot be assumed to be Gaussian. Thus, the particle filter is better under this condition compared to KF. The position estimation is realized based on the particle filter either with at least three detected accurate ranges or with the detected accurate ranges and range differences.

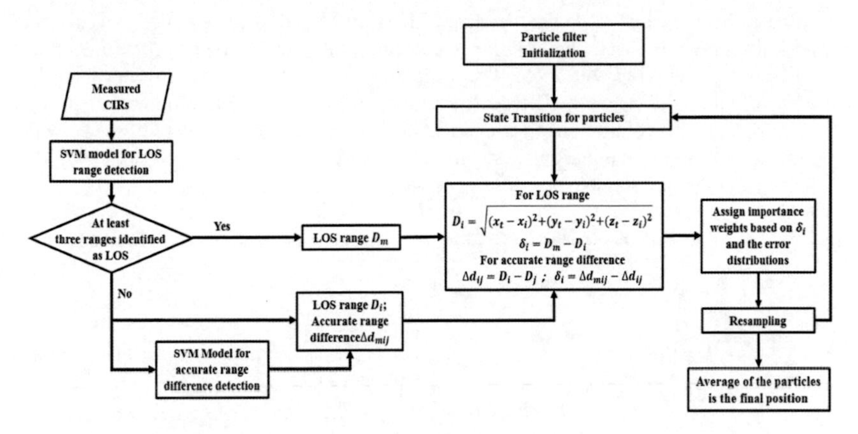

Figure 6.4: Flowchart of the proposed TOA/TDOA particle filter with the LOS range and accurate range difference identification

The particle filter was already introduced in the previous chapters. The process of the LOS range selection based on SVM is the same as in the last chapter. The accurate range difference identification will be presented in this chapter. The flowchart of the proposed TOA/TDOA particle filter with the LOS range and accurate range difference identification is illustrated in Figure 6.4.

6.2 Accurate Range Difference Detection

Although the range selection process is the same as in the last chapter, the optimal feature combination and the effective signal length of the CIR based on SVM are different due to the changed environment. The optimal feature combination for the Bosch Changsha plant comprises the distance, energy, kurtosis, PFTD and mean excess delay. The effective signal length of the CIR ranges from 4 to 70. The P_{LOS} and P_{NLOS} with this combination are 93.78% and 95.12%. Once the localization environment changes, the signal propagation path changes so dramatically that the characteristics of the CIRs are also different. Thus, the optimal feature combinations and the effective signal length of the CIR can change accordingly. Principally, as soon as the localization environment changes, the optimization process (similar to in last chapter) must be evaluated again.

If at least three ranges are detected as accurate, the position estimation is realized with these ranges. However, if less than three ranges are accurate, the accurate range differences, that are calculated with the inaccurate ranges, need to be identified and used for localization. Similar to the LOS range detection, the accurate range difference selection can be achieved with SVM. The testing environment here is the Bosch Changsha plant. Figure 6.5 shows the CIR pairs for range differences in four different situations: the accurate range difference with two LOS CIRs; the accurate range difference with two NLOS CIRs; the inaccurate range difference with one NLOS CIR and one LOS CIR; and the inaccurate range difference with two NLOS CIRs.

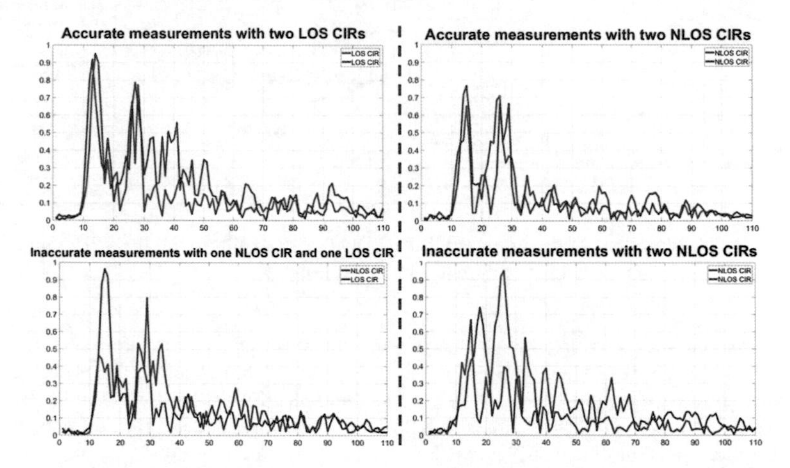

Figure 6.5: Range difference error distribution for the third and fifth BSs

Since the LOS ranges are identified with the first kind of SVM model and are used directly for position estimation, the second kind of SVM model only needs to classify the accurate and inaccurate range differences, that are calculated based on the NLOS CIRs.

Each range difference needs to be computed using two ranges, which means that one CIR pair (containing two CIRs) determines one range difference. Thus, the features, which can be used for accurate range difference detection, are twice so many as these used for LOS range detection. Furthermore, in the used UWB system, it can be observed that the difference of the CIR pair of the inaccurate range differences is slightly different compared to the difference of the CIR pair for accurate range differences, as shown in Fig. 6.6. For the calculation of the CIR difference, it is very important to ensure that the start point of the first path is the first point for the two CIRs. Similar to LOS range detection, the energy, maximal amplitude, standard deviation, and kurtosis of the difference of the CIR pair can also be used as features. Furthermore, the dynamic time warping and the cosine similarity of the CIR pair are also considered.

Together with these features extracted from the difference of the CIR pair, the following features for each CIR in the pair can be used: 1) distance; 2) energy; 3) maximal amplitude of the CIRs; 4) standard deviation; 5) kurtosis; 6) form factor; 7) crest factor; 8) skewness; 9) rise time to the maximal amplitude; 10) PFTD; 11) mean excess delay; 12) RMS delay spread.

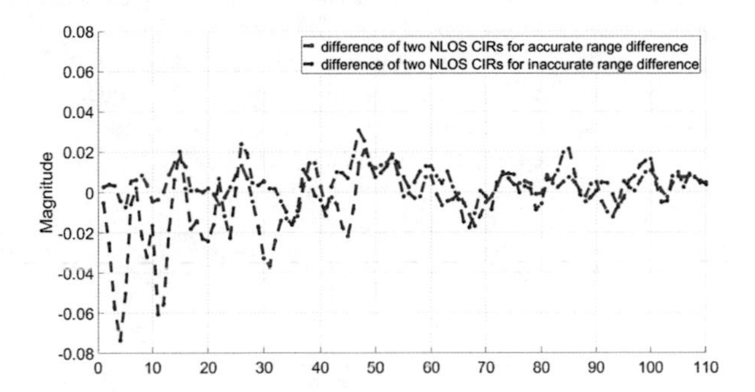

Figure 6.6: Difference between two CIRs for accurate and inaccurate range difference

In total 30 different features can be extracted. However, neither all features nor all received CIR sampling points are needed. In the experiments, as the number of used features and CIR sampling points decreases, the identification accuracy can be improved and the computer load decreases. Thus, the used feature combination and the effective length of the CIR sampling points need to be optimized. CIRs with 63 different lengths of the sampling points are investigated: starting from 1 to 110, 100 ... 30; 4 to 110, 100 ... 30; ... 19 to 110, 100 ... 30. Thus, for the SVM model, in total $63 * (2^{30}\text{-}1)$ different feature combinations with different CIR lengths can be used for classification. The best solution needs to be determined among these $63 * (2^{30}\text{-}1)$ possibilities. Unlike the LOS range detection, a theoretical analysis to determine the best possible solution is difficult to achieve. Hence, the genetic algorithm is used to solve this optimization problem. Because compared to other traditional methods, this algorithm is more efficient and faster. Furthermore, thanks to the crossover and mutation operator, this algorithm is more immune to being trapped in a local optimum.

Genetic Algorithm

The genetic algorithm is used to find the optimal solution to a problem, and it is inspired by natural evolution. A gene, that contains '1's and '0's, is used to represent one solution to the problem. The genes used in this paper consist of two parts. The first part contains 63 possibilities, from '000000' to '111111'. '000000' and '111111' represent the same CIR length. Thus, the first part indicates 63 possibilities for different CIR lengths. The second part of a gene contains 30 '1's and '0's, which represents the different feature combinations. '0' means that the feature is not used, whereas '1' means the feature is used. All '1's and all '0's mean that all the features are used and not used, respectively. All $63 * (2^{30} - 1)$ possible solutions are represented by the genes. The flowchart of this algorithm is shown in Fig. 6.7. The algorithm starts with a random generation of 30 different genes. During the evaluation phase, SVM is used to train the training data with the extracted features from different CIR lengths. The classification accuracy is evaluated with the testing data. A predefined amount of iteration steps is used to terminate the loop. In the selection phase, genes with good accuracy have better chances to be selected and passed to the next generation. The accuracy performance is evaluated using the parameter P.

$$P = P_{LOS} + 2.5P_{NLOS} = \frac{CL}{AL} + 2.5\frac{CN}{AN} \tag{6.5}$$

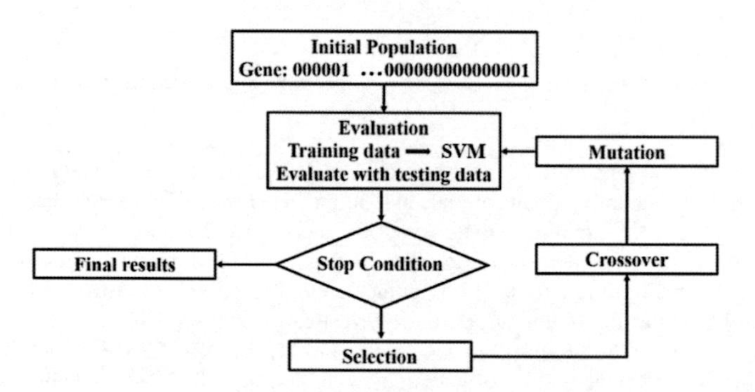

Figure 6.7: Flowchart of the genetic algorithm

Unlike LOS range detection, a larger weight is given to the inaccurate range difference detection to ensure that as few inaccurate measurements are selected as possible. One of the main reasons for this is that during the testing, it is found that the NLOS range detection accuracy is much higher than the inaccurate range difference detection. In addition, as soon as enough accurate measurements are available, further increasing the amount of used accurate data can no longer improve the accuracy. However, if one of the inaccurate measurements is used for calculation, the localization accuracy can be dramatically decreased. With the help of mutation and crossover, the genetic diversity is maintained and the existing genes are combined or changed into new genes. After the mutation and crossover phase, the new feature combinations from different CIR lengths are fed into the evaluation step again. If the predefined number of iterations is reached, the iteration process stops. More details about genetic algorithms can be found in [AK16], [Kan03].

6.3 Test Results

Seven BSs and one MS are used for testing. The Decawave chips are integrated in the BSs and MS. The range measurements and CIRs are collected. At the beginning, all data collected from different BSs are combined into one training dataset. However, the detection accuracy for the accurate range difference is not as good as expected. This is mainly because the NLOS CIRs received by different BSs can be very different even under the same conditions. Based on the experimental results, if a separate training model is built for each BS pair, the accuracy can be improved. Thus, for each BS pair, a separate model is built. Since seven BSs are used here, a total of 21 models are built. For NLOS range identification, a total of seven SVM models are trained.

Furthermore, while training the SVM model for accurate range difference detection, the best classification accuracy is achieved with different CIR lengths and different feature combinations for different BS pairs. The CIR length and feature combination with the best classification accuracy for the first BS is determined using the genetic algorithm. The classification accuracy performance with the same CIR length and feature combination is used to train the dataset from the remaining BSs. Then the average of the classification accuracy is calculated. The same process is repeated for the other BSs. For real time applications, the length of the CIR needs to be the same and predefined for all BSs; thus the CIR length and the feature combination with the best average classification accuracy is selected for SVM to train the dataset for simplification.

A total of 50,000 CIR samples are used as training data for each BS, and 20,000 samples for each BS are utilized as testing data in SVM. The training data and testing data are collected in static and dynamic conditions. In the static condition, the MS is fixed in more than 50 positions. In the dynamic condition, the MS is attached on the shoulder of a person and this person walks along a predefined trajectory. The CIRs and measured ranges are collected. The real ranges are measured based on the Bosch GLM 100 C Professional laser measure, the floor map and cameras. The CIRs are labeled as LOS or NLOS according to the difference between the real ranges and the measured ranges. These CIRs are used to train the NLOS range detection SVM model. The CIR pairs are labeled as accurate and inaccurate according to the difference between the real range differences and the calculated range differences, and they are used as an input vector for training the accurate range difference SVM model.

After the training models are built, the accurate ranges or range differences can be selected based on these models. The particle filter is used for position estimation with the selected measurements.

SVM Classification Performance

As mentioned earlier in this chapter, the optimal feature combination for the accurate range identification in the Bosch Changsha plant comprises the distance, energy, kurtosis, PFTD and mean excess delay. The effective signal length of the CIR ranges from 4 to 70. The P_{LOS} and P_{NLOS} with this combination after the SVM parameter tuning are 93.78% and 95.12%. Due to the change of the localization environment, the signal propagation path changes so dramatically that the characteristics of the CIRs also change. Thus, the optimal feature combinations and the effective signal length of the CIR needs to be changed accordingly. Principally, as soon as the localization environment changes, the optimization process (similar to the last chapter) need to be evaluated again.

Table 6.1: Accuracy of accurate range difference detection with different feature combinations extracted from different CIR lengths for the third BS

	1 to 40	**1 to 60**	**1 to 80**	**1 to 110**	**7 to 60**	**7 to 80**
3*2	53.12%	61.71%	62.19%	63.54%	67.41%	69.81%
5*2	53.75%	65.14%	70.45%	69.43%	61.51%	70.12%
10*2	60.54%	63.46%	67.87%	69.72%	70.08%	63.19%
5*2,a	59.73%	67.98%	70.13%	70.83%	69.16%	71.93%
6*2,a,b	61.33%	63.76%	69.72%	71.73%	70.57%	69.85%

Table 6.2: Accuracy of inaccurate range difference detection with different feature combinations extracted from different CIR lengths for the third BS

	1 to 40	**1 to 60**	**1 to 80**	**1 to 110**	**7 to 60**	**7 to 80**
3*2	60.18%	62.79%	65.19%	62.25%	59.49%	60.59%
5*2	60.86%	67.71%	69.79%	73.71%	71.23%	70.84%
10*2	70.12%	77.19%	78.17%	76.91%	70.12%	69.79%
5*2,a	69.75%	78.71%	73.84%	73.55%	79.89%	77.75%
6*2,a,b	66.49%	79.12%	78.99%	70.13%	71.87%	75.33%

Similar to the accurate range detection, two parameters need to be tuned to improve the range difference classification accuracy: C and Gamma. For different feature combinations, the C and Gamma are different. Table 6.1 and 6.2 list the accuracy for accurate and inaccurate range difference detection with different feature combinations extracted from different CIR lengths for the third BS after the optimization of the C and Gamma in SVM. The features used in these tables are chosen in the following order: directly from each CIR in the CIR pair, 1) distance; 2) energy; 3) maximal amplitude of the CIRs; 4) standard deviation; 5) kurtosis; 6) form factor; 7) rise time to the maximal amplitude; 8) PFTD; 9) mean excess delay; 10) RMS delay spread; from the difference of the CIR, a) energy b) maximal amplitude c) standard deviation d) kurtosis. The first column indicates the used feature combination; for example, '3*2' means that the first three features are extracted from both CIRs for the BS pair, while '3*2, a' means that the first three features are extracted from both CIRs in the BS pair and the energy of the difference between the CIRs is also used. The first row indicates the used length of the CIR. No specific relationship between the classification accuracy, the number of used features and the CIR length can be observed. Thus multiple local optimal solutions are expected. The main reason to use the genetic algorithm is that it is more suitable for optimization problems with multiple local optima compared to other algorithms.

During the testing, it is found that the optimal feature combination and the effective signal length of the CIR varies for different BSs. However, it is not convenient to use different feature combinations and different effective signal lengths of the CIR for different BSs. Thus, the feature combination and the effective signal length of the CIR with the best average accuracy for these BSs are determined and used in the application. The best overall accuracy is achieved with the following feature combination: RMS delay spread-, received signal energy-, distance between BS and MS-, crest factor for each CIR in the CIR pair, maximal amplitude and standard deviation extracted from the difference between the CIRs. The effective signal length of the CIR ranges from 4 to 80. The average accuracy of accurate and inaccurate range difference identification for seven BSs are respectively 75.79% and 82.89%.

Since the NLOS range identification accuracy is very high, the noise distribution after the NLOS detection can still be taken as Gaussian distribution. The range error distribution for the sixth BS after combining the errors for the correctly detected accurate ranges and incorrectly identified inaccurate ranges is presented in Figure 6.8. However, the range difference error distribution can not be assumed to be Gaussian due to the not very promising identification accuracy. One of the range difference error distributions after combining the errors for the correctly detected accurate range differences and incorrectly identified inaccurate range differences is presented in Figure. 6.9.

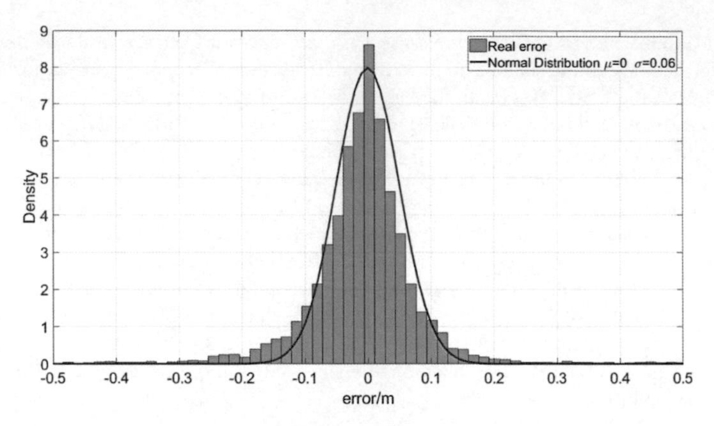

Figure 6.8: Range error distribution for the sixth BS

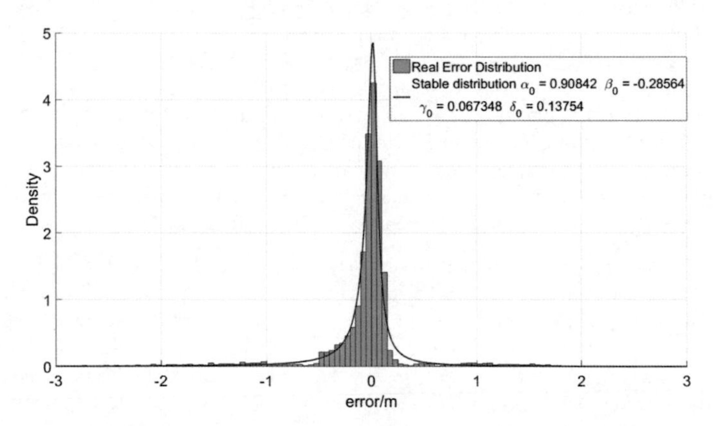

Figure 6.9: Range difference error distribution for the third and fifth BS

Localization Performance

During the field test, the seven used BSs are fixed in the following positions: BS1=[0.3 -0.3], BS2=[13 -0.3], BS3=[13, -4.3], BS4=[13 -8.6], BS5=[0.3 -8.6], BS6=[0.3 -4.3] and BS7=[6.8 -2.3]. The particle filter is used for the localization. The pseudocode of the particle filter algorithm with accurate ranges and accurate range differences identification is shown in Algorithm 2.

Algorithm 2 Particle Filter with NLOS detection Pseudo Code

1: Initialization
2: **for** t=2 $\rightarrow$ total time steps **do**
3: **for** j=1 $\rightarrow$ used number of particles **do**
4: $X_t = AX_{t-1} + W$
5: Accurate ranges detection (D_{ij})
6: **if** n=> 3 $\rightarrow$ number of detected accurate ranges **then**
7: $D_{ij} = \sqrt{(x_{jt} - x_i)^2 + (y_{jt} - y_i)^2 + (z_{jt} - z_i)^2}$
8: $\delta_{ij} = D_{mi} - D_{ij}$
9: $w_j = \prod_{1 \leq i \leq n} f(\delta_{ij})$
10: **else**
11: $D_{ij} = \sqrt{(x_{jt} - x_i)^2 + (y_{jt} - y_i)^2 + (z_{jt} - z_i)^2}$
12: $\delta_{ij} = D_{mi} - D_{ij}$
13: Accurate range differences detection (D_{fgj})
14: k is the number of detected accurate range difference detection
15: $D_{gj} = \sqrt{(x_{jt} - x_g)^2 + (y_{jt} - y_g)^2 + (z_{jt} - z_g)^2}$
16: $D_{fj} = \sqrt{(x_{jt} - x_f)^2 + (y_{jt} - y_f)^2 + (z_{jt} - z_f)^2}$
17: $D_{fgj} = D_{fj} - D_{gj}$
18: $\delta_{fgj} = D_{mfg} - D_{fgj}$
19: $w_j = \prod_{i \leq 2} f_{range}(\delta_{ij}) \prod_{1 \leq i \leq k} f_{rangedifference}(\delta_{ifj}; \alpha, \beta, \gamma, \delta)$
20: **end if**
21: **end for**
22: Normalize weights
23: Resample
24: $X_t = \frac{X_{t1} + X_{t2} + \cdots + X_{tN}}{N}$
25: **end for**

Figure 6.10 shows the localization results based on the normal particle filter, the LOS range detection based particle filter and the LOS range and accurate range difference detection based TOA/TDOA combined particle filter. The red line is the predefined trajectory. The black points are the localization results with the TOA based on the particle filter with NLOS range detection. As shown in this figure, compared to the other two approaches, the position estimation accuracy with the NLOS range detection is slightly better. However, the NLOS error can not be reduced in many places, especially around the upper-right corner. The main reason for this is that during the test, usually a maximum of two or even no BSs are identified as LOS. Most of the current approaches assume that enough LOS BSs are available during the localization process, and they only detect whether the BSs are under NLOS. There is no further discussion about the cases in which only two or even none LOS BSs are detected. However, in harsh industrial areas, where the signal propagation is blocked most of the time, the accuracy improvement can not be achieved by solely detecting whether the signal is under NLOS due to the lack of LOS BSs. Compared to the existing identification approaches, the proposed approach further identifies the accurate range difference, which is calculate based on the NLOS ranges with roughly the same errors. With the detected accurate range differences, accurate position estimation with less than three LOS BSs can be achieved. The position estimation results of the proposed TOA/TDOA particle filter with accurate range and range difference identification are shown as blue points in Figure. 6.10. Very promising accuracy improvements can be observed with the proposed approach in the upper-right corner. In that corner, most of the time only the third BS is under clear LOS; the other BSs are blocked by the machines. Due to the workers or the forklift frequently passing the gate, the second BS is also under NLOS. This is why the other algorithm can not have accurate localization results. With the help of the selected accurate range differences, accurate position estimation can still be achieved. Overall, the localization accuracy is improved with the proposed approach based on the accurate range and range difference identification.

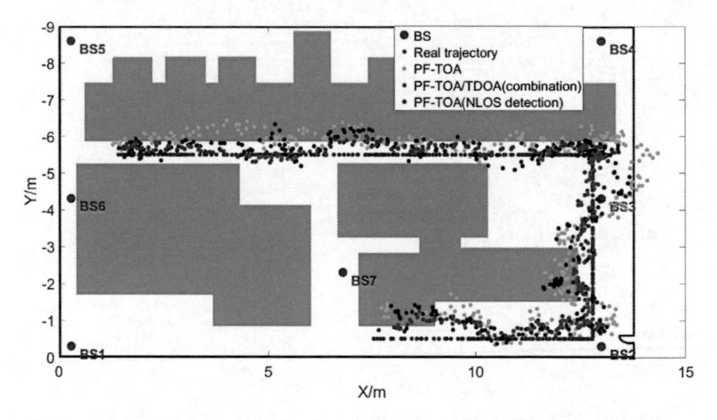

Figure 6.10: Field test localization results based on different approaches

6.4 Summary

This chapter focused on UWB position estimation in harsh industrial environments. An accurate range and range difference identification based TOA/TDOA combination approach is proposed to improve the localization accuracy. Two SVM models are trained. The first one is used to select the accurate ranges. If at least three ranges are detected as accurate, then the localization can be realized with these ranges. Otherwise, the range differences need to be calculated with the inaccurate ranges. The second SVM model is used to select the accurate range differences. These accurate range differences and ranges are used for position estimation. The particle filter is used to realize the accurate range and range difference identification based TOA/TDOA combination approach. The position estimation with the proposed TOA/TDOA combination approach shows better accuracy compared to the other approaches in the Bosch Changsha plant.

7 UWB NLOS Detection and Mitigation Based on IMU

The 9-axis IMU system contains a 3-axis gyroscope, a 3-axis accelerometer and a 3-axis magnetometer. Unlike wireless based indoor localization, the NLOS conditions do not have any influence on IMU based localization and furthermore, the system is infrastructure-free. However, the IMU system suffers from drift errors due to biases. The UWB system does not have a drift problem, but the NLOS leads to inaccurate position estimation. The fusion of UWB and IMU can utilize the advantages of both systems to improve the localization accuracy. As mentioned in the previous chapters, many papers have discussed the fusion of IMU and UWB. In this study, the IMU measurements are used to identify and mitigate the UWB NLOS errors. The fusion system can be used for both TOA and TDOA based on UWB localization. With the help of the IMU measurements, the accurate ranges for TOA or the accurate range differences for TDOA can be determined based on the triangle inequality theorem.

Figure 7.1: Size of BNO055 [bno]

The IMU used in the project is the Bosch developed BNO055, which integrates a triaxial 14-bit accelerometer, a triaxial 16-bit gyroscope with a range of ± 2000 degrees per second, and a triaxial geomagnetic sensor [hs]. With its very small size (5.2 x 3.8 x 1.1 mm^3, as shown in Figure 7.1), this sensor can still provide relatively accurate acceleration, angular velocity and magnetic field strength information [hs]. Like the other IMU products, the system noise of the gyroscope and accelerometer is mainly caused by the system white noise, bias instability, temperature affected bias, the scale factors, and alignments, as presented in Table 7.1.

Table 7.1: Error Type [Woo07],[WSFI]

Type of Error	Description
Offset	Constant bias in output signal
Thermo-mechanical white noise	Normally assumed as normally distributed with zero mean
Instability of the bias	Can be modeled as random walk
Temperature	As the temperature changes, the bias changes
Calibration	Scale factors, alignments, and linearity (deterministic errors)

The acceleration in x direction can be represented as [WSFI]:

$$a_{xm} = a_x + c_{ax} + b_{ax} + w_{ax} \tag{7.1}$$

where a_{xm} is the measured acceleration in x direction. a_x is the real acceleration, c_{ax} is the constant offset (bias) for the acceleration, b_{ax} is the moving bias (caused by the instability of the bias and the changes in the temperature) and w_{ax} is the white noise.

Similar to the acceleration in x direction, the acceleration in the y, z direction and the gyroscope output can be modeled with a similar model, e.g.:

$$g_{xm} = g_x + c_{gx} + b_{gx} + w_{gx} \tag{7.2}$$

where g_{xm} is the measured gyroscope output in x direction, g_x is the real gyroscope output, c_{gx} is the constant offset (bias) for the gyroscope output, b_{gx} is the moving bias (caused by the instability of the bias and the changes in the temperature) and w_{gx} is the white noise.

The error performance descriptions for BNO055 can be found in [bno]. The greatest challenge in IMU based localization is reducing the drift error caused by biases. Figure 7.2 and Figure 7.3 present the drift error of the measurements from the gyroscope and accelerometer in a static state. It can be observed in these figures, that the distances calculated by integrating the accelerations from IMU are only accurate in a very short time. In the proposed method, the IMU measurements are mainly used to detect whether the UWB measurements are accurate. The update rate of UWB is 20 Hz. For every 0.05 second, the localization is updated based on the UWB system. The prediction with IMU data for every 0.05 second can be assumed to be accurate.

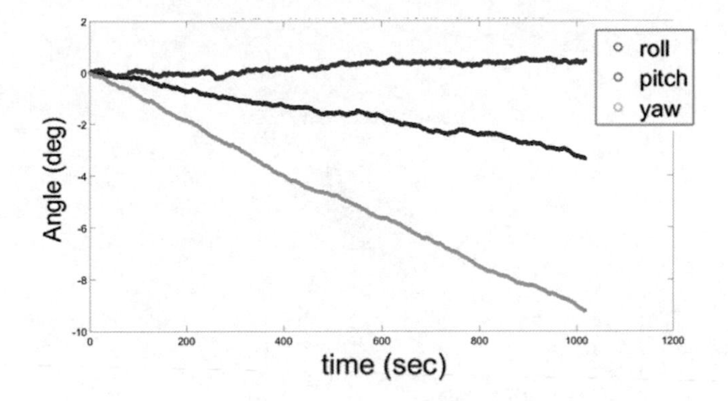

Figure 7.2: Drift error of the angle in a static state in 16 minutes

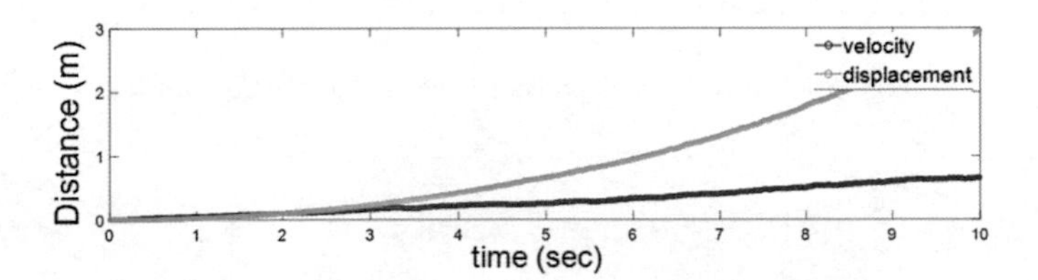

Figure 7.3: Drift error of the position in a static state in 10 seconds

As discussed in the previous section, accurate position estimation is achieved with accurate range measurements for TOA based localization, and with accurate range differences in TDOA based localization. The IMU measurements can be used to select the accurate range and range difference. This chapter shows the UWB localization accuracy improvement approaches with accurate measurement selection for TOA and TDOA. The angular velocity and the acceleration from IMU are measured with respect to the IMU body-fixed coordinate. However, the UWB localization is realized in the global coordinate, as presented in Figure 7.4. To fuse the two systems, the first step is to transform the measured acceleration from the IMU body-fixed coordinates to the global coordinates.

The Euler angle or quaternion can be used for coordinate transformation. Due to the singularities problem for Euler angle, quaternion is used for the transformation. Quaternion can be directly calculated by the integration of the angular velocity. However, the results drift over time due to bias. Based on the gravity vector measured by the accelerometer and magnetic strength, the quaternion can be determined without the

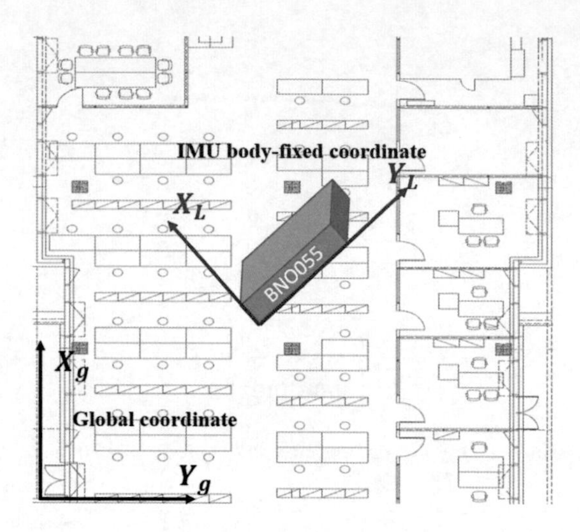

Figure 7.4: IMU body-fixed coordinate and global coordinate

drift problem. However, the results are not so smooth and the accuracy can be guaranteed only if the accelerometer is motionless. The advantages of both techniques can be utilized by fusion all three sensors, while the disadvantages can be avoided. An extended Kalman filter (EKF) approach is proposed to combine these sensors in [YB06]. The state vector in the EKF is 7-D (3-D angular rates and 4-D quaternion). Despite its good accuracy, due to its high non-linearity, the computational requirements for such a filter are too high; therefore, for real time applications, this approach is not the best choice. The Euler angle is calculated with a simplified Kalman filter in [TLL⁺13]. In this project, a similar simplified Kalman filter is used for the quaternion calculation. Instead of seven states, the state vector only contains the 4-D quaternion, which makes the filter completely linear. The third order low-pass Butterworth filter is applied as a pre-filter. Then, the quaternion is calculated based on the acceleration and magnetic strength with the Gauss-Newton algorithm and used afterwards as measurement input for the Kalman filter. The prediction in the filter is realized with the angular velocities. Furthermore, the values in the measurement noise covariance matrix R_k are increased in case the total angular rate exceeds a threshold, since in a motion state, the predicted value is more reliable. Conversely, the values decrease, since in a static state the results calculated by the Gauss-Newton algorithm are more accurate, because they do not suffer from drift problems.

Quaternion Calculation Based on the Gauss-Newton Algorithm

In the first step, the quaternion is calculated with the Gauss-Newton algorithm based on the acceleration and magnetic strength [Com]. With q_4 being the real component, the quaternion can be written as:

$$\boldsymbol{q}(t) = [q_1\ q_2\ q_3\ q_4] \tag{7.3}$$

The acceleration and magnetic strength can be written in the following matrix:

$$\boldsymbol{m}(t) = \begin{pmatrix} a_x(t) \\ a_y(t) \\ a_z(t) \\ M_x(t) \\ M_y(t) \\ M_z(t) \end{pmatrix} \tag{7.4}$$

Assuming that the body-fixed coordinates and the global coordinates are the same at the initial position, the initial state of $\boldsymbol{m}_t$ is:

$$\boldsymbol{m}_0 = \begin{pmatrix} 0 \\ 0 \\ 9.81 \\ M_{x0} \\ M_{y0} \\ M_{z0} \end{pmatrix} \tag{7.5}$$

The coordinate transformation from any frame to the global frame and the error $\boldsymbol{\xi}$ can be expressed as:

$$\boldsymbol{m}_0 = \boldsymbol{M}\boldsymbol{m} \tag{7.6}$$

$$\boldsymbol{\xi} = \boldsymbol{m}_0 - \boldsymbol{M}\boldsymbol{m} \tag{7.7}$$

where

$$\boldsymbol{M} = \begin{pmatrix} \boldsymbol{R} & 0 \\ 0 & \boldsymbol{R} \end{pmatrix} \tag{7.8}$$

$$\boldsymbol{R} = \begin{pmatrix} q_4^2 + q_1^2 - q_2^2 - q_3^2 & 2(q_1q_2 - q_3q_4) & 2(q_1q_3 + q_2q_4) \\ 2(q_1q_2 + q_3q_4) & q_4^2 + q_2^2 - q_1^2 - q_3^2 & 2(q_3q_2 - q_1q_4) \\ 2(q_1q_3 - q_2q_4) & 2(q_3q_2 + q_1q_4) & q_4^2 + q_3^2 - q_2^2 - q_1^2 \end{pmatrix} \tag{7.9}$$

Based on the Gauss-Newton algorithm, the calculation of the quaternion at time t can be realized with the following equations:

$$\boldsymbol{q}(t) = \boldsymbol{q}(t-1) - [J_{t-1}^T(\boldsymbol{q}(t-1))J_{t-1}(\boldsymbol{q}(t-1))]^{-1}J_{t-1}(\boldsymbol{q}(t-1))\boldsymbol{\xi}(\boldsymbol{q}(t-1)) \tag{7.10}$$

where J_{t-1} is the Jacobian matrix:

$$J_{t-1}(\boldsymbol{q}(t-1)) = \frac{\partial \boldsymbol{\xi}}{\partial \boldsymbol{q}(t-1)}$$
$$= -[\frac{\partial \boldsymbol{M}}{\partial q_1}\boldsymbol{m}(t-1) \quad \frac{\partial \boldsymbol{M}}{\partial q_2}\boldsymbol{m}(t-1) \quad \frac{\partial \boldsymbol{M}}{\partial q_3}\boldsymbol{m}(t-1) \quad \frac{\partial \boldsymbol{M}}{\partial q_4}\boldsymbol{m}(t-1)] \tag{7.11}$$

To reach the optimum point at each time step, the iteration process is applied. The following equation represent the ith iteration at time t:

$$\boldsymbol{q}_{i+1}(t) = \boldsymbol{q}_i(t) - [J_t^T(\boldsymbol{q}_i(t))J_t(\boldsymbol{q}_i(t))]^{-1}J_t(\boldsymbol{q}_i(t))\boldsymbol{\xi}(\boldsymbol{q}_i(t)) \tag{7.12}$$

The iteration stops in case the length difference between $\boldsymbol{q}_{i+1}(t)$ and $\boldsymbol{q}_i(t)$ is smaller than 0.0001 or i is equal to 8. The calculated quaternion is used as a measurement input for the Kalman filter.

Quaternion Update with the KF

The fusion of the calculated quaternion and the gyroscope measurements is realized with the KF to improve the quaternion accuracy. The state vector in the KF is the quaternion:

$$\boldsymbol{X}_t = \begin{pmatrix} q_4 \\ q_1 \\ q_2 \\ q_3 \end{pmatrix} \tag{7.13}$$

The quaternion derivative can be updated with the measurements from the gyroscope [Mad]:

$$\dot{\boldsymbol{q}}(t) = \frac{1}{2}\boldsymbol{q}(t-1) \otimes \boldsymbol{\omega}(t) \tag{7.14}$$

where

$$\boldsymbol{\omega}(t) = \begin{pmatrix} 0 \\ \omega_x(t) \\ \omega_y(t) \\ \omega_z(t) \end{pmatrix} \tag{7.15}$$

$\otimes$ is the quaternion product. Assuming that $\boldsymbol{a}$ and $\boldsymbol{b}$ are vectors, then [Mad]:

$$\boldsymbol{a} \otimes \boldsymbol{b} = [a_1 \;\; a_2 \;\; a_3 \;\; a_4] \otimes [b_1 \;\; b_2 \;\; b_3 \;\; b_4] = \begin{pmatrix} a_1b_1 - a_2b_2 - a_3b_3 - a_4b_4 \\ a_1b_2 + a_2b_1 + a_3b_4 - a_4b_3 \\ a_1b_3 - a_2b_4 + a_3b_1 + a_4b_2 \\ a_1b_4 + a_2b_3 - a_3b_2 + a_4b_1 \end{pmatrix}^T \tag{7.16}$$

The prediction in the KF can be achieved with the following equation:

$$\hat{\boldsymbol{X}}_t = \boldsymbol{X}_{t-1} + \dot{\boldsymbol{q}}(t-1)T = A\boldsymbol{X}_{t-1} \tag{7.17}$$

where A is the state transition model:

$$A = \begin{pmatrix} 1 & -0.5\omega_x(t)T & -0.5\omega_y(t)T & -0.5\omega_z(t)T \\ 0.5\omega_x(t)T & 1 & 0.5\omega_z(t)T & -0.5\omega_y(t)T \\ 0.5\omega_y(t)T & -0.5\omega_z(t)T & 1 & 0.5\omega_x(t)T \\ 0.5\omega_z(t)T & 0.5\omega_y(t)T & -0.5\omega_x(t)T & 1 \end{pmatrix} \tag{7.18}$$

and T is the time difference between the current time t and the previous time $t-1$.

The calculated quaternion based on the Gauss-Newton algorithm with the acceleration and magnetic strength is used as a measurement input for the Kalman filter. The update process is the same as the one described in the previous chapter. The structured flowchart for the whole calculation process is presented in Figure 7.5.

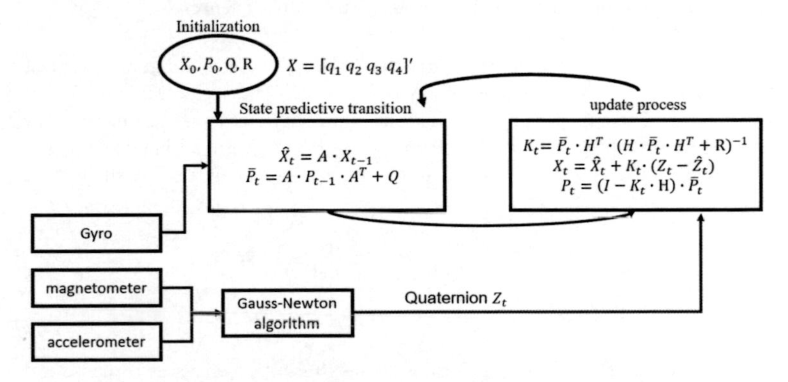

Figure 7.5: Structured flowchart for the KF

After the quaternion is obtained, the acceleration can be transformed from the IMU body-fixed coordinates to the global coordinates with the following equation [CH12], [TKA16]:

$$\boldsymbol{a}_g = \boldsymbol{q}_t \boldsymbol{a}_l (\boldsymbol{q}_t)' \tag{7.19}$$

where $\boldsymbol{a}_g$ is the acceleration in global coordinates and $\boldsymbol{a}_l$ is the acceleration in the IMU body-fixed coordinates.

7.1 TOA Based UWB/IMU Fusion System

Compared to CIR based NLOS identification, UWB/IMU fusion based NLOS mitigation processes less data, so the computation load is lighter. Instead of processing the long CIR vector and calculating many features, the IMU only provides nine elements in the vector. The particle filter is more suitable than the KF for this problem, since it is more stable and the system is non-linear. However, the calculation load of the particle filter is heavier than that of the KF. The application of the UWB/IMU fusion system in this project needs to realize fast response works with few delays. Thus, the calculation load has to be kept at a minimum. Considering the stability and the computation complexity, the IEKF is a good compensation solution. The fusion algorithm can be realized in three steps. Firstly, the accelerations are transformed into global coordinates. Secondly, the accurate ranges are selected based on the moving distance prediction with the IMU measurements. This can be realized in the predictive state in the IEKF. Thirdly, the position of the MS is updated with the detected accurate ranges in the IEKF.

NLOS Identification Based on the Triangle Inequality Theorem

The triangle inequality theorem states that the length of any side of a triangle is always shorter than the sum of the lengths of any two sides. LOS range detection is achieved based on this theorem. As demonstrated in Figure 7.6, the ranges at time points t-1 and t, together with the distance from time points t-1 to t, form a triangle. According to the triangle inequality theorem, the following inequations must be satisfied:

$$\Delta S + D_{mi,t-1} - D_{mi,t} > 0 \tag{7.20}$$

$$\Delta S + D_{mi,t} - D_{mi,t-1} > 0 \tag{7.21}$$

$$D_{mi,t-1} + D_{mi,t} - \Delta S > 0 \tag{7.22}$$

Since the update rate is 20 Hz, the distance change ΔS in such a short time is almost

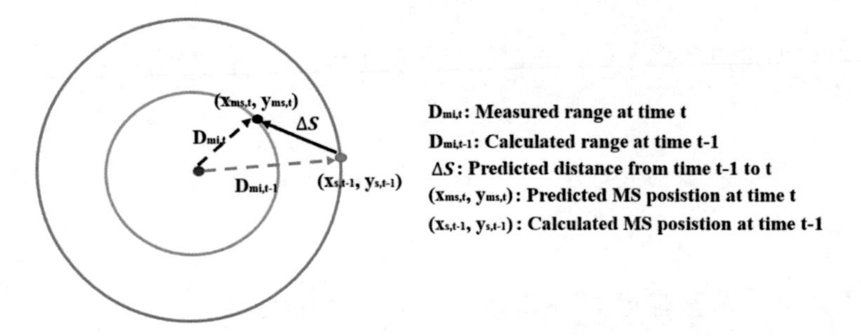

Figure 7.6: Geometry illustration of triangle inequality theorem in TOA based localization

always smaller than the sum of two ranges. Thus, equation 7.22 always holds. The following equation can be obtained after combining equation 7.20 and equation 7.21:

$$\Delta S - ||D_{mi,t} - D_{mi,t-1}|| > 0 \tag{7.23}$$

The predicted moving distance can be calculated in the prediction state of the IEKF:

$$\hat{X}_t = \begin{pmatrix} x_{ms,t} \\ y_{ms,t} \\ v_{xmt} \\ v_{ymt} \end{pmatrix} = AX_{t-1} + Bu \tag{7.24}$$

The moving distance ΔS is:

$$\Delta S = \sqrt{(x_{ms,t} - x_{s,t-1})^2 + (y_{ms,t} - y_{s,t-1})^2} \tag{7.25}$$

The range at time t-1 can be calculated after the update process in IEKF.

$$D_{mi,t-1} = \sqrt{(x_{s,t-1} - x_i)^2 + (y_{s,t-1} - y_i)^2} \tag{7.26}$$

where (x_i, y_i) is the position of the ith BS.

The real measured range $D_{rmi,t}$ is in 3D. Since the heights of the MS and BS are fixed, the range in 2D can be computed as follows:

$$D_{mi,t} = \sqrt{D_{rmi,t}^2 - (z_s - z_i)^2} \tag{7.27}$$

where z_s is the height of the MS and z_i is the height of the BS.

In theory, if equation 7.23 does hold, the measured range could be accurate. In reality, the errors need to be considered:

$$\Delta S - ||D_{mi,t} - D_{mi,t-1}|| = \Delta S - ||D_{i,t} + \varepsilon_{i,t} + b_{i,t} - D_{i,t-1} - \Delta\mu_{i,t-1}|| > 0 \tag{7.28}$$

where $D_{i,t}$ and $D_{i,t-1}$ are the real distances at time t and $t-1$, $\varepsilon_{i,t}$ is the system random noise error, $b_{i,t}$ is the NLOS error under NLOS conditions, $\Delta\mu_{i,t-1}$ is the calculated range error at the previous time point, $\varepsilon_{i,t} + b_{i,t}$ is defined as the measurement error at the current time, and $\varepsilon_{i,t} + b_{i,t} - \Delta\mu_{i,t-1}$ is the total error.

If the range measurements with the smallest measurement error can be selected for calculation, accurate position estimation can be achieved. However, the corresponding range measurements with the smallest total error, that satisfy equation 7.23, are selected. Depending on the previous position estimation error and the current measurement error, the range measurements with the smallest total error might not be those with the smallest measurement error. If the previous position estimation error is small, the calculated range error at the previous time point is also small. Four possibilities have to be discussed [ZLW18c]:

(1) With a small position estimation error (small range error) at the previous time point and a small measurement error, the accurate range will be selected as expected.

(2) With a small position estimation error and a large measurement error, the inaccurate range will not be selected as expected.

(3) With a large position estimation error for the previous time point and a small measurement error, the accurate measurement will not be selected. In this case, the position estimation can still be accurate if three accurate range measurements can be chosen, since the number of used range measurements is not important when enough accurate range measurements are selected.

(4) With a large position estimation error in the previous time point and a large measurement error, two possibilities need to be discussed:

First, if the estimation error has the opposite sign to the measurement error, the measurement will not be selected as expected.

Second, if the estimation error happens to have the same sign as the measurement error, then the measurement might unfortunately be detected as accurate and the position accuracy will be affected by this measurement. This almost never happens during the testing. One solution to this problem is to use the IMU measurements to predict the ranges, and to then compare the predicted ranges with the current range measurements. The measurements that are considered as inaccurate show significant differences compared to the predictive ranges. However, this case never occurs during the simulation and real field tests in this study. For simplification purposes, the comparison process is ignored here.

Due to the existing bias and signal random noise in the acceleration measurements, the estimated maximum error of the acceleration needs to be set as a threshold. This means that only if the left side of equation 7.29 is larger than the threshold, this measurement can be considered accurate. This can be expressed as follows: only if equation 7.29 holds, can the corresponding range measurement be selected.

$$\Delta S - ||D_{mi,t} - D_{mi,t-1}|| > Threshold \qquad (7.29)$$

After NLOS range detection, the accurate ranges are selected and used as measurements in the update process in the IEKF, which was presented in Chapter 3. Same as in the previous chapters, the Gaussian distributions are used to describe the noise distributions. To mitigate the NLOS error, the measurement noise covariance matrix need to be changed once the ranges are detected as inaccurate. Assuming ξ_i is the standard deviation of the measurement noise for the i_{th} BS under LOS, a parameter de_i is used to tune the noise in the measurement noise covariance matrix. R_t can be represented as:

$$R_t = \begin{pmatrix} de_1\xi_1\xi_1 & 0 & 0 & \dots & 0 \\ 0 & de_2\xi_2\xi_2 & 0 & \dots & 0 \\ \vdots & \vdots & \vdots & \dots & \vdots \\ 0 & 0 & 0 & \dots & de_N\xi_N\xi_N \end{pmatrix} \tag{7.30}$$

If the measurement is detected as accurate, the corresponding de_i is set to be a small value. Otherwise, the value is set to be large, so that the influence of the corresponding measurement can almost be ignored. The UWB/IMU localization with the NLOS identification process is presented in Figure 7.7.

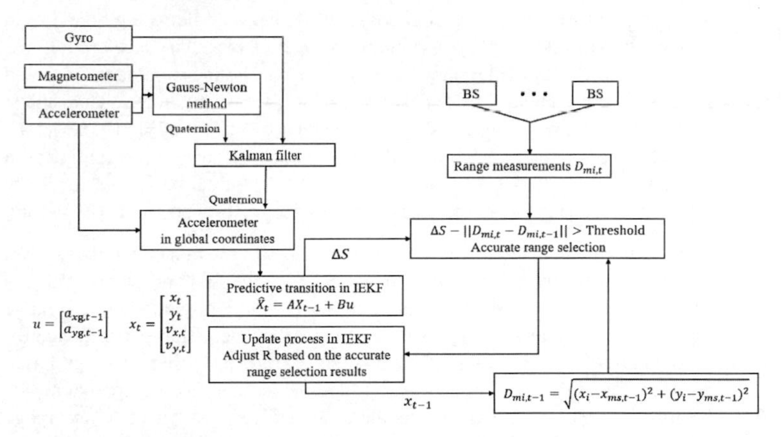

Figure 7.7: Structured flowchart for UWB/IMU fusion with the accurate range selection approach

Simulation Results

The NLOS detection accuracy based on the proposed approach is very hard to be determined in real field tests. This is because the evaluation needs to be done in a dynamic process, which means that the UWB/IMU system need to be kept moving. It is not easy to determine the NLOS/LOS condition in a dynamic process with 100% accuracy. Hence, the NLOS detection performance of the IMU based approach is evaluated in a simulation. The tightly coupled UWB/IMU method proposed in [HDLS09] is used for comparison. The NLOS outlier detection with this method is based on the following residuals distribution:

$$\delta_{ij,t} = D_{mi,t} - D_{ij,t} \tag{7.31}$$

where $D_{mi,t}$ is the measured range at time t, and $D_{ij,t}$ is the predicted range based on the IMU data. The distribution of $\delta_{ij,t}$ should be a normal distribution in the absence of errors [HDLS09]:

$$\delta_{ij,t} \sim \mathcal{N}(0, C_t P_{t|t-1} C_t^T + R_t)[\text{HDLS09}] \tag{7.32}$$

where C_t is the measurement Jacobian, $P_{t|t-1}$ is the state covariance and R_t is measurement noise covariance matrix. In case the calculated confidence intervals for the individual measurement are violated, the measurement is considered as a NLOS outlier [HDLS09].

In the simulation, the range measurements are corrupted by zero mean Gaussian noise to represent the measurement noise, and some of the measurements are added by a positive value to represent NLOS errors caused by positive bias. The acceleration was corrupted by non-zero mean Gaussian noise to represent the measurement noise and bias. The amount of the under NLOS BSs are used as the variable factors to evaluate all approaches. In the simulation, the BSs are located at x1=[0 0], x2=[5 0], x3=[5 7], x4=[0 7], x5=[-0.5 3], and x6=[5.5 3] and the predefined trajectory is a rectangle with vertex position [1.2 1.3], [3.4 1.3], [3.4 4.6], [1.2 4.6]. The final position error is the average of the difference between the measurements and the true positions [ZLW18c].

Two different cases are considered during the simulation. In the first one, only the fifth and sixth BSs are under NLOS. In the second case, only the first and second BSs are under LOS. For the NLOS BSs, 100 positive biases are randomly added to the range measurements. Table 7.2 and Table 7.3 present the NLOS detection performance for two different approaches under two different situations: the proposed approach based on the triangle inequality theorem, and the tightly coupled UWB/IMU approach. '1st BS' represents the first BS, 'Total' means the total detected number of NLOS range measurements based on the NLOS identification approaches; and 'Real' means the real number of NLOS range measurements that are detected by the NLOS identification approach.

Table 7.2: NLOS identification performance with fifth and sixth BSs under NLOS

	1st BS	2nd BS	3rd BS	4th BS	5th BS	6th BS
	Total/Real	∼	∼	∼	∼	∼
Triangle inequality theorem based	11/0	3/0	9/0	17/0	117/95	112/89
Tightly coupled UWB/IMU approach based	5/0	2/0	6/0	10/0	80/70	87/69

Table 7.3: NLOS identification performance with third, fourth, fifth, and sixth BSs under NLOS

	1st BS	2nd BS	3rd BS	4th BS	5th BS	6th BS
	Total/Real	∼	∼	∼	∼	∼
Triangle inequality theorem based	17/0	9/0	119/92	128/90	125/89	121/88
Tightly coupled UWB/IMU approach based	12/0	3/0	89/65	92/71	98/75	89/69

Overall, the NLOS detection possibility for the triangle inequality theorem approach is above 90%, while for the tightly coupled UWB/IMU approach it is 70%. The NLOS identification accuracy is the key factor for accurate position estimation. The position errors using these two approaches are presented in Figure 7.8. It can be observed that the localization accuracy using the triangle inequality theorem method is better.

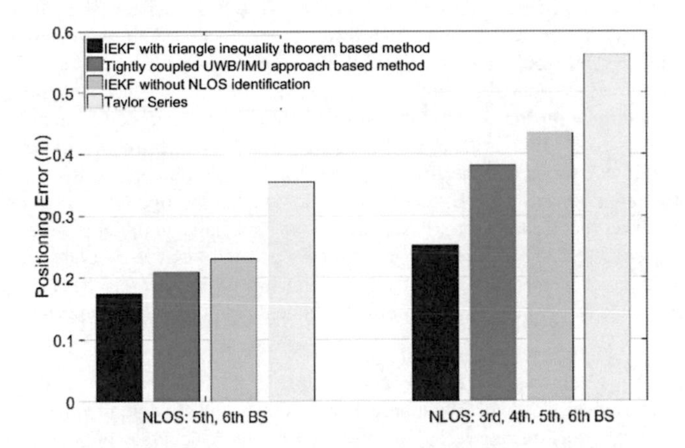

Figure 7.8: Simulation results for different approaches in two different cases

However, the erroneous detection rate of the triangle inequality theorem approach is higher than that of the tightly coupled approach. Furthermore, the more BSs are under

NLOS, the higher the erroneous detection rate is. E.g. in the first case, for the fifth BS with the proposed approach, 22 accurate range measurements (117-95) are detected as inaccurate, while in the second case, this number is 36 (125-89). One of the possibilities to reduce the erroneous detection rate for the proposed approach is to reduce the threshold value. However, with a smaller threshold, the NLOS detection rate decreases. Compared to the erroneous detection rate, the NLOS detection rate is more critical for position estimation accuracy, because as soon as the LOS BSs are used for calculation, the accuracy can be guaranteed. The number of LOS BSs is secondary. For example, in the 2D scenario, if five BSs are under LOS, in the worst case, only three BSs are identified as LOS and used for the calculation, the accuracy could still be guaranteed, even if only one BS is identified as LOS, the proper BSs could still be chosen, because the erroneously detected LOS BSs still have the smallest residues and the corresponding de_i is still small. Figure 7.8 shows the triangle inequality theorem based approach has the best accuracy performance, since the range measurements under NLOS have less influence on the calculation. The approaches with NLOS identification have better accuracy than those without NLOS detection.

Field Test Results

The localization performance is evaluated in the Bosch Shanghai office environment. Figure 7.9 shows the test results using different approaches. The red rectangle in the figure is the reference trajectory. Overall, the approaches with NLOS detection have better accuracy compared to those without NLOS detection. Thanks to the accurate NLOS identification, both the triangle inequality theorem NLOS detection based IEKF approach and the IEKF with tightly coupled UWB/IMU method are more accurate than the Taylor series and the IEKF without NLOS detection, especially in the right bottom corner and near the BS5. However, sometimes the tightly coupled approach fails to select the accurate range measurements. Especially in the upper-right corner, this approach selects inaccurate measurements and ignores the accurate ones; thus, the accuracy is worse than that of the triangle inequality theorem based approach. Overall, the triangle inequality theorem NLOS detection based IEKF approach has the best accuracy.

Figure 7.10 presents the localization performance of the IEKF with triangle inequality theorem based NLOS detection approach and the PF with triangle inequality theorem based NLOS detection approach. The position estimation accuracy of the PF approach is slightly better in the bottom-left corner. However, the calculation time in Matlab for this PF based approach is 2.088056 seconds, while for the IEKF it is 0.090820 seconds. The IEKF is much faster than the PF based approach. Thus, due to the fast response requirement of the application, the IEKF with triangle inequality theorem based NLOS detection approach is more suitable.

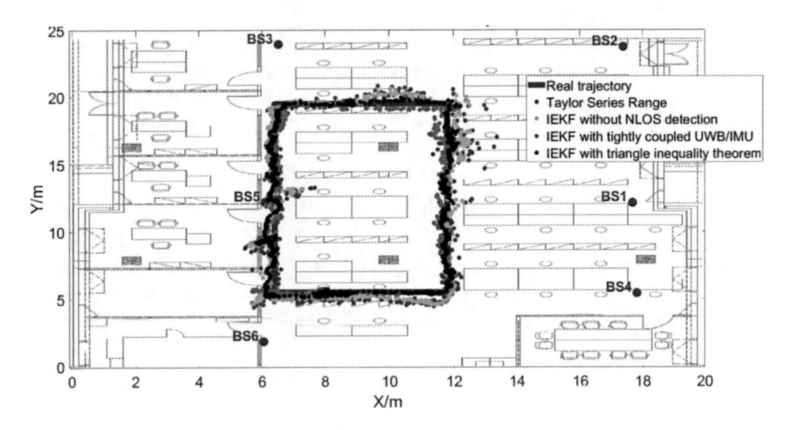

Figure 7.9: Field test results using different approaches

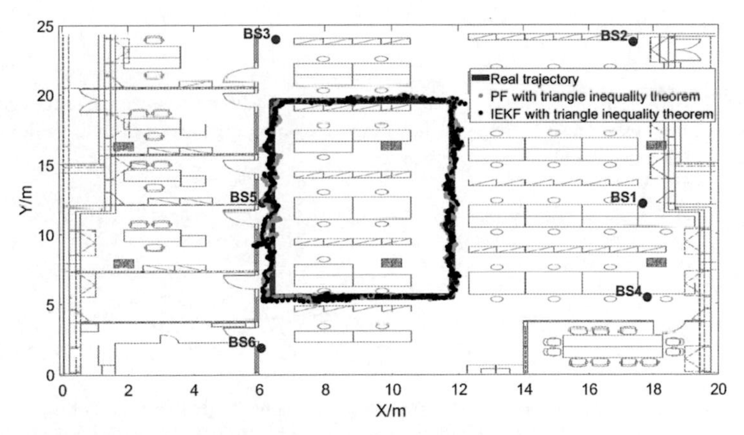

Figure 7.10: Position estimation performance for the IEKF and PF with triangle inequality theorem based NLOS detection approaches

7.2 TDOA Based UWB/IMU Fusion System

Unlike the UWB/IMU fusion system with triangle inequality theorem based NLOS detection for TOA approach, the accurate position estimation with TDOA based UWB and IMU fusion system is achieved with accurate range differences. Similarly, accurate range differences can also be identified with the triangle inequality theorem.

Accurate Range Differences Selection Based on the Triangle Inequality Theorem

As described above, the ranges at time points t-1 and t, together with the distance ΔS from time points t-1 to t, form a triangle. For two BSs, two triangles can be formed, as shown in Figure 7.11. The following equations can be derived according to the triangle inequality theorem:

$$\Delta S + D_{mi,t-1} - D_{mi,t} > 0 \tag{7.33}$$

$$\Delta S + D_{mi,t} - D_{mi,t-1} > 0 \tag{7.34}$$

$$D_{mi,t-1} + D_{mi,t} - \Delta S > 0 \tag{7.35}$$

$$\Delta S + D_{mj,t-1} - D_{mj,t} > 0 \tag{7.36}$$

$$\Delta S + D_{mj,t} - D_{mj,t-1} > 0 \tag{7.37}$$

$$D_{mj,t-1} + D_{mj,t} - \Delta S > 0 \tag{7.38}$$

As discussed earlier, equations 7.35 and 7.38 almost always hold, since the ranges are always larger than the short time distance change. Combining the rest of the equations, the following two equations can be obtained:

$$\begin{aligned} &2\Delta S + D_{mi,t} - D_{mj,t} + D_{mj,t-1} - D_{mi,t-1} \\ &= 2\Delta S + d_{mij,t} - d_{mij,t-1} > 0 \end{aligned} \tag{7.39}$$

$$\begin{aligned} &2\Delta S + D_{mi,t-1} - D_{mj,t-1} + D_{mj,t} - D_{mi,t} \\ &= 2\Delta S + d_{mij,t-1} - d_{mij,t} > 0 \end{aligned} \tag{7.40}$$

where ΔS can be predicted based on IMU measurements in the KF prediction phase, $d_{mij,t}$ is the range difference at the current time point directly measured by UWB, and $d_{mij,t-1}$ is the calculated range difference at the previous time step. Combining equations 7.39 and 7.40 yields:

$$2\Delta S + ||d_{mij,t} - d_{mij,t-1}|| > 0 \tag{7.41}$$

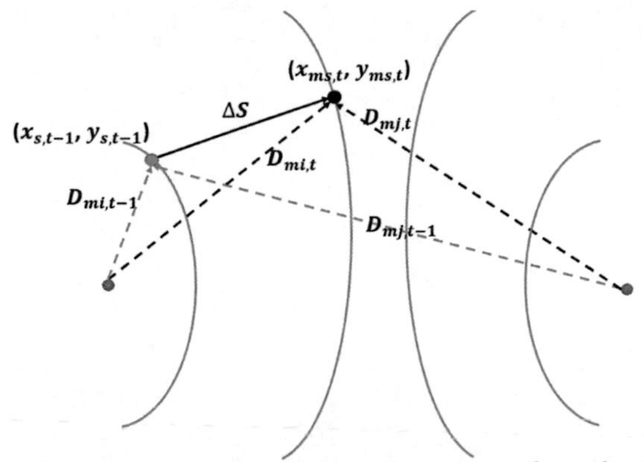

$D_{mi,t-1} - D_{mj,t-1}$: Calculated range difference for the ith and j^{th} BSs at time t-1
$D_{mi,t} - D_{mj,t}$: Measured range difference for the ith and j^{th} BSs at time t
ΔS: Predicted distance from time t-1 to t
$(x_{s,t-1}, y_{s,t-1})$: Calculated MS posistion at time t-1
$(x_{ms,t}, y_{ms,t})$: Predicted MS posistion at time t

Figure 7.11: Geometric presentation of the triangle inequality theorem for TDOA

The predicted moving distance can be calculated in the prediction state of the IEKF:

$$\hat{X}_t = \begin{pmatrix} x_{ms,t} \\ y_{ms,t} \\ v_{xmt} \\ v_{ymt} \end{pmatrix} = AX_{t-1} + Bu \tag{7.42}$$

The moving distance ΔS is:

$$\Delta S = \sqrt{(x_{ms,t} - x_{s,t-1})^2 + (y_{ms,t} - y_{s,t-1})^2} \tag{7.43}$$

The range difference at time t-1 can be calculated after the update process in the IEKF.

$$d_{mij,t-1} = \sqrt{(x_{s,t-1} - x_i)^2 + (y_{s,t-1} - y_i)^2} - \sqrt{(x_{s,t-1} - x_j)^2 + (y_{s,t-1} - y_j)^2} \tag{7.44}$$

where (x_i, y_i) is the position of the ith BS and (x_j, y_j) is the position of the jth BS.

The real measured range difference $d_{rmij,t}$ is in 3D, while the triangle inequality theorem is valid in 2D. Based on the fact that the heights of the MS and BS are fixed, the range difference can be simplified in 2D as follows. In the 3D case:

$$d_{rmij,t} = \sqrt{(x_{s,t} - x_i)^2 + (y_{s,t} - y_i)^2 + (z_s - z_i)^2}$$

$$- \sqrt{(x_{s,t} - x_j)^2 + (y_{s,t} - y_j)^2 + (z_s - z_j)^2} \tag{7.45}$$

where z_s is the height of the MS and z_i is the height of the BS.

In the 2D case:

$$d_{mij,t} = \sqrt{(x_{s,t} - x_i)^2 + (y_{s,t} - y_i)^2} - \sqrt{(x_{s,t} - x_j)^2 + (y_{s,t} - y_j)^2} \tag{7.46}$$

Combining equations 7.45 and 7.46 produces:

$$\begin{aligned} & d^2_{mij,t} - d^2_{rmij,t} = \\ & 2\sqrt{(x_{s,t} - x_i)^2 + (y_{s,t} - y_i)^2 + (z_s - z_i)^2}\sqrt{(x_{s,t} - x_j)^2 + (y_{s,t} - y_j)^2 + (z_s - z_j)^2} \\ & - 2\sqrt{(x_{s,t} - x_i)^2 + (y_{s,t} - y_i)^2}\sqrt{(x_{s,t} - x_j)^2 + (y_{s,t} - y_j)^2} - 2(z_s - z_i)^2 \end{aligned} \tag{7.47}$$

There are two possibilities to calculate $d_{mij,t}$:

1) Since the update rate is 20 Hz, the position change from t-1 to t is very small, and it can be assumed that: $(x_{s,t}, y_{s,t}) \approx (x_{s,t-1}, y_{s,t-1})$. Thus:

$$\begin{aligned} d^2_{mij,t} = {} & d^2_{rmij,t} + \\ & 2\sqrt{(x_{s,t-1} - x_i)^2 + (y_{s,t-1} - y_i)^2 + (z_s - z_i)^2} \\ & \sqrt{(x_{s,t-1} - x_j)^2 + (y_{s,t-1} - y_j)^2 + (z_s - z_j)^2} \\ & - 2\sqrt{(x_{s,t-1} - x_i)^2 + (y_{s,t-1} - y_i)^2}\sqrt{(x_{s,t-1} - x_j)^2 + (y_{s,t-1} - y_j)^2} \\ & - 2(z_s - z_i)^2 \end{aligned} \tag{7.48}$$

2) Under the condition that $(z_s - z_j)$ is very small, (in other words, the height difference between the BS and MS is very small), it can be assumed that:

$$d_{mij,t} \approx d_{rmij,t} \tag{7.49}$$

Theoretically if equation 7.41 does hold, the measured range difference could be accurate. In reality, the errors need to be considered:

$$\begin{aligned} & 2\Delta S + ||d_{mij,t} - d_{mij,t-1}|| = \\ & \quad 2\Delta S - ||d_{mij,t} - d_{mij,t-1} + \varepsilon_{i,t} + b_{i,t} - \varepsilon_{j,t} - b_{j,t} - \Delta\mu_{i,t-1}|| > 0 \end{aligned} \tag{7.50}$$

where $\varepsilon_{i,t}$ is the system random noise error for the ith BS, $b_{i,t}$ is the NLOS error for the ith BS under NLOS condition, $\Delta\mu_{i,t-1}$ is the calculated range difference error at the previous time point, $\varepsilon_{i,t} + b_{i,t} - \varepsilon_{j,t} - b_{j,t}$ is defined as the measurement error at the current time point, and $\varepsilon_{i,t} + b_{i,t} - \varepsilon_{j,t} - b_{j,t} - \Delta\mu_{i,t-1}$ is the total error.

Only the measurements that satisfy equation 7.50 are selected for further calculation. This means that the corresponding range difference can only be selected if the total error is small. However, the goal is to select the range differences with the smallest measurement error at the current time for further calculation. Similarly as discussed before, based on the previous position estimation error and the current measurement error, four possibilities need to be discussed [ZLW18b]:

(1) If the position estimation error at the previous time point is small, the calculated range difference error is small. With a large measurement error, the measurement will not be selected as expected.

(2) The calculated range difference error is small. With a small measurement error, the measurement will be selected as expected.

(3) If the range difference error in the previous time point is large, but the measurement error is small, the measurement will not be selected. If enough accurate range differences are selected, the number of used range differences is not important. Thus, even if some accurate measurements are not chosen, accurate position estimation is still possible.

(4) The range difference error in the previous time point is large and so is the measurement error. Again, there are two possibilities:

First, if the range difference error has the opposite sign to the measurement error, the measurement will not be selected as expected.

Second, if the range difference error happens to have the same sign as the measurement error, the measurement will unfortunately be selected and have a negative influence on the estimation. This rarely happens and the predictive position based on IMU can be used to predict the range difference and compared with the current measurement range difference. The measurements that show a significant difference compared to the predictive estimation can be taken as inaccurate ones. However, this case almost never occurs during the simulation and field test in this study. For simplification purposes, the comparing process can be ignored.

Since the acceleration contains bias and signal noise, the estimated maximum error value of the acceleration is set as a threshold. This means that only if the left side of equation 7.51 is larger than the threshold, the conclusion that this measurement is accurate can be made. This can be expressed as follows: if equation 7.51 holds, the corresponding range difference measurement can be selected.

$$2\Delta S + ||d_{mij,t} - d_{mij,t-1}|| > Threshold \tag{7.51}$$

After the accurate range differences are selected, these data are used as measurements in the update process in the IEKF, which was covered in Chapter 3. To mitigate the NLOS error, the measurement noise covariance matrix need to be changed once the range differences are detected as inaccurate. Assuming that $\xi_i j$ is the measurement noise for the range difference from the i_{th} and j_{th} BSs, a parameter de_ij is used to tune the noise in the measurement noise covariance matrix. R_t can be represented as:

$$R_t = \begin{pmatrix} de_{12}\xi_{12}\xi_{12} & 0 & 0 & 0 & \dots & 0 \\ 0 & de_{13}\xi_{13}\xi_{13} & 0 & 0 & \dots & 0 \\ \vdots & \vdots & \vdots & \vdots & \dots & \vdots \\ 0 & 0 & de_{1N}\xi_{1N}\xi_{1N} & 0 & \dots & 0 \\ 0 & 0 & 0 & de_{23}\xi_{23}\xi_{23} & \dots & \vdots \\ \vdots & \vdots & \vdots & \vdots & \dots & \vdots \\ 0 & 0 & 0 & 0 & \dots & de_{(N-1)N}\xi_{(N-1)N}\xi_{(N-1)N} \end{pmatrix} \tag{7.52}$$

If the measurement is detected as accurate, the corresponding de_ij is set to be a small value, and otherwise the value is set to be large, so that the influence of the corresponding measurement can almost be ignored. The UWB/IMU localization process with accurate range difference selection is presented in Figure 7.12.

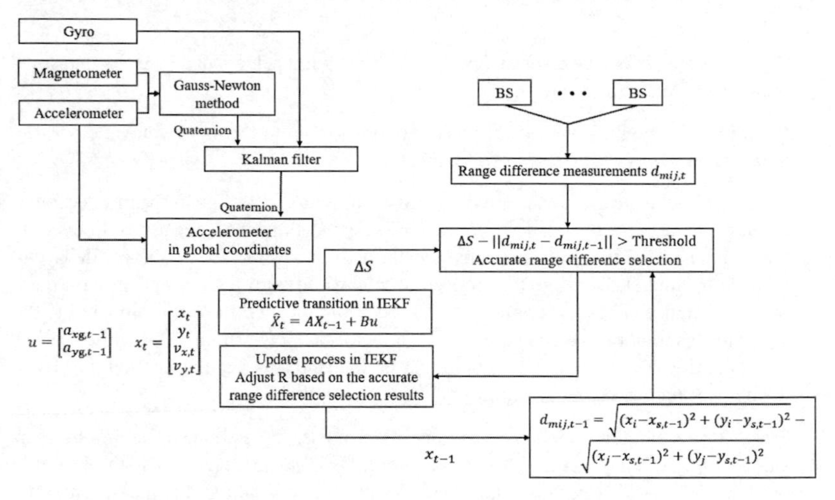

Figure 7.12: Structured flowchart for the UWB/IMU fusion approach with range difference selection

Localization Performance

The localization performances of the IEKF with accurate range difference selection, the normal IEKF, the Taylor series, and the PF are evaluated in the Bosch Shanghai office environment. Figure 7.13 presents the position estimation results of the different approaches. The red rectangle in the figure is the reference trajectory. The position estimation results of the IEKF approach with accurate range difference selection are better than those of the other methods. Especially on the left side of the trajectory, thanks to the selection of the accurate range difference, the localization performance is clearly better. However, on the bottom right, the method fails to detect the inaccurate range differences. Thus, no improvement can be observed in this part.

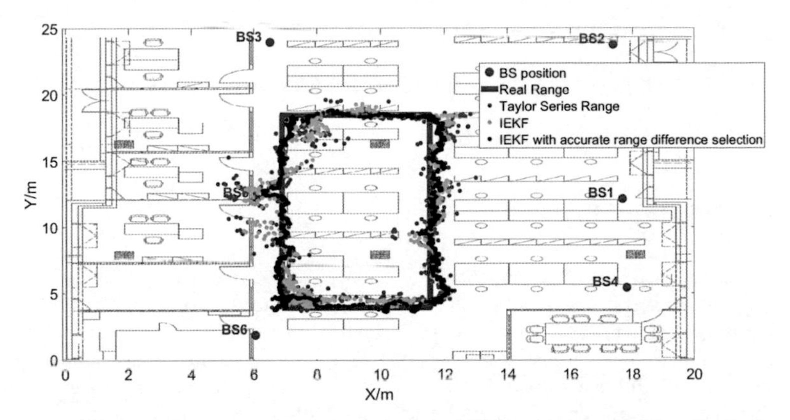

Figure 7.13: Localization performance using the IEKF without selection, the Taylor series and the IEKF with accurate range difference selection

Figure 7.14 shows the position estimation results based on the IEKF/PF with accurate range selection approach. The position estimation accuracy of the PF approach is slightly better in the bottom left corner. However, the calculation time in Matlab for this approach is 5.519828 seconds, and for the IEKF it is 0.139587 seconds. The IEKF is much faster than the PF based approach. Thus, for fast response applications, the IEKF with triangle inequality theorem based accurate range difference approach is more suitable.

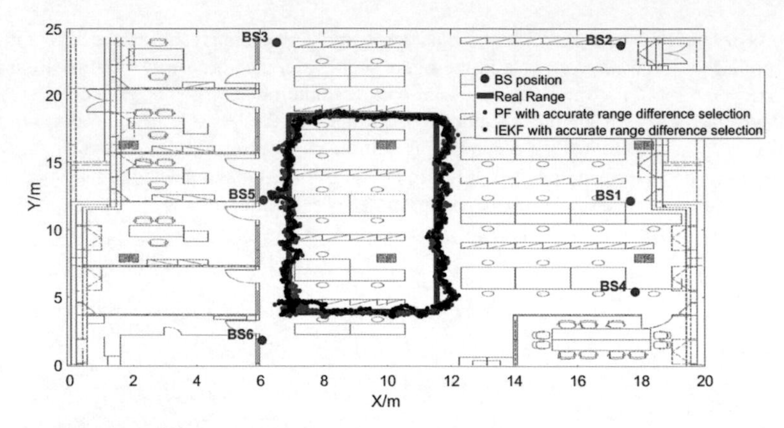

Figure 7.14: Position estimation based on the IEKF/PF with accurate range selection approach

7.3 Summary

In this chapter, the accurate range/range difference measurements are selected with the proposed triangle inequality theorem based approach. The fusion of UWB and IMU is realized with the IEKF. Compared to CIR based NLOS identification, the fusion system is more suitable for fast response applications. Furthermore, the localization performances of different approaches are evaluated. The proposed accurate range/range difference measurements detection based IEKF with the help of IMU data is more accurate compared to the other algorithms.

8 Conclusions

This thesis is based on the Bosch "Real-time Safety Virtual Positioning" (RSVP) project, which aims to replace the fixed installation of safeguards with an indoor localization based safety system to increase productivity and flexibility. The first Chapter provided an overview of existing indoor localization systems. The UWB was selected as the most suitable system for the RSVP project after comparing these systems based on the following factors: accuracy, update rate, system capacity, system complexity, coverage and cost. The accuracy of the localization system is one of the most important factors for the RSVP application. The inaccurate position estimation for the UWB system is mainly caused by NLOS errors. Thus, in this thesis, different algorithms were proposed to improve UWB based indoor localization under NLOS conditions.

A detailed description of the UWB system was given, and the factors that influence UWB localization accuracy were summarized. These factors are the antenna, installation of the BSs, time synchronization, localization algorithms, NLOS/LOS identification and filter algorithms. The relationship between CIR and accurate/inaccurate measurements was theoretically explained in different cases. These relationships were also evaluated based on the collected CIRs in real office environments. In addition, the thesis investigated UWB range measurement error under clear LOS/multipath LOS, ignorable NLOS blockage and non-ignorable NLOS blockage. It was found that the ranges under LOS are accurate, and even the measurements under glass, table, and chair NLOS blockages are trustable. This kind of NLOS blockage with accurate ranges was defined in this paper as soft NLOS (SNLOS). In contrast, hard NLOS (HNLOS) blockages lead to inaccurate range measurements. In this thesis, NLOS identification was used to detect HNLOS blockages.

As one of the most effective approaches to NLOS mitigation, NLOS identification methods are discussed in many papers. This thesis presented an overview of these methods, and then compared them based on identification accuracy, engineering feasibility, and so forth. After the comparison, it was determined that CIR based and IMU based NLOS identification are two of the best approaches. Hence, different approaches were developed for CIR based and IMU based NLOS identification to improve the UWB accuracy. For NLOS mitigation, the EKF, IEKF, and PF were the focus of this thesis; they were described and compared.

The error distribution in the Bosch Shanghai office was determined to be a stable distribution. If only three BSs are used for localization, the localization accuracy can be improved with a properly defined error distribution. However, the NLOS error is still severe. A better solution to further improve accuracy is to add redundant BSs and use only the selected accurate ranges based on NLOS identification for further calculation. The SVM algorithm was used for CIR based NLOS detection. Five different feature groups were divided based on distance, CIR shape, time, multipath richness, and power related features. The reasons why these features can be used for identification were explained. The feature combination, the used CIR length, and the SVM parameters were optimized to improve the identification accuracy during the field test. The localization accuracy with NLOS identification was dramatically improved compared to the other methods in the Bosch Shanghai office environment.

Furthermore, the thesis discussed UWB localization in harsh industrial environments. NLOS conditions occur more frequently in harsh industrial environments than in office environments. The localization environment was in the Bosch Changsha plant. It happened frequently that less than two ranges were measured under LOS during the field tests. The improvement in the accuracy with NLOS identification was limited due to the lack of accurate ranges. Thus, a TOA/TDOA combination approach based on accurate range and range difference identification was proposed to further improve the position estimation accuracy. Two SVM models were trained. The first one was used to select the accurate ranges. If at least three ranges were detected as accurate, then the localization was realized with the TOA approach. Otherwise, the range differences had to be calculated with the inaccurate ranges. The second SVM model was used to select the accurate range differences. These accurate range differences and accurate ranges were used for position estimation. The particle filter was utilized to realize TOA/TDOA combination approach based on accurate range and range difference identification. The position estimation with this TOA/TDOA combination approach showed better accuracy than the other approaches in the Bosch Changsha plant.

Finally, a new UWB/IMU fusion approach was proposed. The fusion system can be used for both TOA and TDOA based on UWB localization. With the help of the IMU measurements, the accurate ranges for TOA or the accurate range differences for TDOA can be determined based on the triangle inequality theorem. The localization performance of the UWB/IMU fusion system was evaluated in the Bosch Shanghai office. The localization accuracy was improved with the proposed method compared to those without fusion with IMU.

Bibliography

[AK16] Bakirtzis Anastasios and Spyros Kazarlis. *"Genetic algorithms." Advanced Solutions in Power Systems: HVDC, FACTS, and Artificial Intelligence: HVDC, FACTS, and Artificial Intelligence*. 2016.

[AKP15] A. Alamin, H. M. Khalid, and J. C. . Peng. Power system state estimation based on iterative extended kalman filtering and bad data detection using normalized residual test. In *2015 IEEE Power and Energy Conference at Illinois (PECI)*, pages 1–5, Feb 2015.

[Bd08] M. Bouet and A. L. dos Santos. Rfid tags: Positioning principles and localization techniques. In *2008 1st IFIP Wireless Days*, pages 1–5, Nov 2008.

[BN18] H. Benzerrouk and A. V. Nebylov. Robust imu/uwb integration for indoor pedestrian navigation. In *2018 25th Saint Petersburg International Conference on Integrated Navigation Systems (ICINS)*, pages 1–5, May 2018.

[bno] https://www.bosch-sensortec.com/bst/products/all_products/bno055.

[BWH11] A. C. Braun, U. Weidner, and S. Hinz. Support vector machines, import vector machines and relevance vector machines for hyperspectral classification - a comparison. In *2011 3rd Workshop on Hyperspectral Image and Signal Processing: Evolution in Remote Sensing (WHISPERS)*, pages 1–4, June 2011.

[CCGZ06] M. Crepaldi, M. R. Casu, M. Graziano, and M. Zamboni. Uwb receiver design and two-way-ranging simulation using vhdl-ams. In *2006 Ph.D. Research in Microelectronics and Electronics*, pages 465–468, June 2006.

[CH12] L. Chen and H. Hu. Imu/gps based pedestrian localization. In *2012 4th Computer Science and Electronic Engineering Conference (CEEC)*, pages 23–28, Sep. 2012.

[CLL18] B. Choi, K. La, and S. Lee. Uwb tdoa/toa measurement system with wireless time synchronization and simultaneous tag and anchor positioning. In *2018 IEEE International Conference on Computational In-*

telligence and Virtual Environments for Measurement Systems and Applications (CIVEMSA), pages 1–6, June 2018.

[CLLW16] Z. Chen, Y. Liu, S. Li, and G. Wang. Study on the multipath propagation characteristics of uwb signal for indoor lab environments. In *2016 IEEE International Conference on Ubiquitous Wireless Broadband (ICUWB)*, pages 1–4, Oct 2016.

[Com] D. Comotti. *ORIENTATION ESTIMATION BASED ON GAUSS-NEWTON METHOD AND IMPLEMENTATION OF A QUATERNION COMPLEMENTARY FILTER*.

[con] https://en.wikipedia.org/wiki/conditional_probability, conditional probability.

[cov] covariance matrix, https://en.wikipedia.org/wiki/covariance_matrix.

[CSW02] R. J. . Cramer, R. A. Scholtz, and M. Z. Win. Evaluation of an ultra-wide-band propagation channel. *IEEE Transactions on Antennas and Propagation*, 50(5):561–570, May 2002.

[Deca] Decawave. aps006 application note, channel effects on communications range and time stamp accuracy in dw1000 based systemst.

[Decb] Decawave. Aps013 application note the implementation of two-way ranging with the dw1000.

[FPS14] C. Forster, M. Pizzoli, and D. Scaramuzza. Svo: Fast semi-direct monocular visual odometry. In *2014 IEEE International Conference on Robotics and Automation (ICRA)*, pages 15–22, May 2014.

[GJD+07] T. Gigl, G. J. M. Janssen, V. Dizdarevic, K. Witrisal, and Z. Irahhauten. Analysis of a uwb indoor positioning system based on received signal strength. In *2007 4th Workshop on Positioning, Navigation and Communication*, pages 97–101, March 2007.

[GRS+17] K. Gururaj, A. K. Rajendra, Y. Song, C. L. Law, and G. Cai. Real-time identification of nlos range measurements for enhanced uwb localization. In *2017 International Conference on Indoor Positioning and Indoor Navigation (IPIN)*, pages 1–7, Sep. 2017.

[GZT08a] Guowei Shen, R. Zetik, and R. S. Thoma. Performance comparison of toa and tdoa based location estimation algorithms in los environment. In *2008 5th Workshop on Positioning, Navigation and Communication*, pages 71–78, March 2008.

[GZT08b] Guowei Shen, R. Zetik, and R. S. Thoma. Performance comparison of toa and tdoa based location estimation algorithms in los environment.

In *2008 5th Workshop on Positioning, Navigation and Communication*, pages 71–78, March 2008.

[HAG17] X. Hou, T. Arslan, and J. Gu. Indoor localization for bluetooth low energy using wavelet and smoothing filter. In *2017 International Conference on Localization and GNSS (ICL-GNSS)*, pages 1–6, June 2017.

[HDLS09] J. D. Hol, F. Dijkstra, H. Luinge, and T. B. Schon. Tightly coupled uwb/imu pose estimation. In *2009 IEEE International Conference on Ultra-Wideband*, pages 688–692, Sep. 2009.

[HIM17] K. A. Horvath, G. Ill, and A. Milankovich. Passive extended double-sided two-way ranging algorithm for uwb positioning. In *2017 Ninth International Conference on Ubiquitous and Future Networks (ICUFN)*, pages 482–487, July 2017.

[Ho] Shirley Ho. Introduction to monte carlo. In *Princeton University*.

[hs] https://www.bosch sensortec.com/bst/products/all_products/bno055. *Bosch*.

[Kan03] M. Kantardzic. *Chapter 10 - Genetic Algorithms. Data Mining: Concepts, Models, Methods, and Algorithms. John Wiley Sons*. 2003.

[Kec01] V. Kecman. *Learning and Soft Computing: Support Vector Machines, Neural Networks, and Fuzzy Logic Models*. MITP, 2001.

[LDBL07] H. Liu, H. Darabi, P. Banerjee, and J. Liu. Survey of wireless indoor positioning techniques and systems. *IEEE Transactions on Systems, Man, and Cybernetics, Part C (Applications and Reviews)*, 37(6):1067–1080, Nov 2007.

[LDM02] V. Lottici, A. D'Andrea, and U. Mengali. Channel estimation for ultra-wideband communications. *IEEE Journal on Selected Areas in Communications*, 20(9):1638–1645, Dec 2002.

[Liu14] J Liu. Survey of wireless based indoor localization technologies. 2014.

[LJHV13] Lorenzo Rubio, Juan Reig, Herman Fernandez, and Vicent M. Rodrigo-Penarrocha. Experimental uwb propagation channel path loss and time-dispersion characterization in a laboratory environment. In *International Journal of Antennas and Propagation*, volume vol. 2013, page 2013, 2013.

[LJXG18] J. Liu, H. Jing, Y. Xiang, and Z. Guan. Simulation research of uwb location algorithm. In *2018 Chinese Control And Decision Conference (CCDC)*, pages 4825–4830, June 2018.

[LLJD15] Q. Li, R. Li, K. Ji, and W. Dai. Kalman filter and its application. In *2015 8th International Conference on Intelligent Networks and Intelligent Systems (ICINIS)*, pages 74–77, Nov 2015.

[LLZP07] Li Li, Li Yu, G. Zhu, and X. Pei. A novel parameter estimation method of alpha-stable distribution based on extreme value. In *2007 Second International Conference on Communications and Networking in China*, pages 621–625, Aug 2007.

[Mad] Sebastian O.H. Madgwick. *An efficient orientation filter for inertial and inertial/magnetic sensor arrays.*

[mai] main diagonal, https://en.wikipedia.org/wiki/main_diagonal.

[Mau12] Rainer Mautz. *Indoor positioning technologies.* PhD thesis, ETH Zurich, 2012.

[Mer13] G & C Merriam. Probability. In *Webster's Revised Unabridged Dictionary*, 1913.

[MGWW10] S. Marano, W. M. Gifford, H. Wymeersch, and M. Z. Win. Nlos identification and mitigation for localization based on uwb experimental data. *IEEE Journal on Selected Areas in Communications*, 28(7):1026–1035, Sep. 2010.

[MMK09] J. Mochnac, S. Marchevsky, and P. Kocan. Bayesian filtering techniques: Kalman and extended kalman filter basics. In *2009 19th International Conference Radioelektronika*, pages 119–122, April 2009.

[MMT15] R. Mur-Artal, J. M. M. Montiel, and J. D. Tardos. Orb-slam: A versatile and accurate monocular slam system. *IEEE Transactions on Robotics*, 31(5):1147–1163, Oct 2015.

[MPS14] L. Mainetti, L. Patrono, and I. Sergi. A survey on indoor positioning systems. In *2014 22nd International Conference on Software, Telecommunications and Computer Networks (SoftCOM)*, pages 111–120, Sep. 2014.

[MSM08] Masaki Yoshino, Shinichiro Haruyama, and Masao Nakagawa. High-accuracy positioning system using visible led lights and image sensor. In *2008 IEEE Radio and Wireless Symposium*, pages 439–442, Jan 2008.

[OCC+17] C. Ou, C. Chao, F. Chang, S. Wang, G. Liu, M. Wu, K. Cho, L. Hwang, and Y. Huan. A zigbee position technique for indoor localization based on proximity learning. In *2017 IEEE International Conference on Mechatronics and Automation (ICMA)*, pages 875–880, Aug 2017.

[PK] Joanna Zietkiewicz Piotr Kozierski, Marcin Lis. resampling in particle filtering - comparison. In *studia z automatyki i informatyki*.

[Pol] David Pollard. Variances and covariances.

[pro] https://en.wikipedia.org/wiki/probability, probability.

[PSWW12] Z. Peng, F. Si, J. Wang, and Y. Wang. Stable distribution and its application in chinese stock market. In *2012 8th International Conference on Natural Computation*, pages 953–956, May 2012.

[SGKJ07] J. Schroeder, S. Galler, K. Kyamakya, and K. Jobmann. Nlos detection algorithms for ultra-wideband localization. In *2007 4th Workshop on Positioning, Navigation and Communication*, pages 159–166, March 2007.

[SH16] B. Silva and G. P. Hancke. Ir-uwb-based non-line-of-sight identification in harsh environments: Principles and challenges. *IEEE Transactions on Industrial Informatics*, 12(3):1188–1195, June 2016.

[SMS12] J. Shen, A. F. Molisch, and J. Salmi. Accurate passive location estimation using toa measurements. *IEEE Transactions on Wireless Communications*, 11(6):2182–2192, June 2012.

[SP12] N. Suenderhauf and P. Protzel. Towards a robust back-end for pose graph slam. In *2012 IEEE International Conference on Robotics and Automation*, pages 1254–1261, May 2012.

[Ter] Gabriel A. Terejanu. Tutorial on monte carlo techniques. In *Department of Computer Science and Engineering, University at Buffalo, Buffalo, NY 14260*.

[TKA16] S. ThoiThoi, K. C. Kodur, and W. Arif. Quaternion based wireless ahrs data transfer using nrf24l01 and hc-05. In *2016 International Conference on Microelectronics, Computing and Communications (MicroCom)*, pages 1–6, Jan 2016.

[TLL+13] K. K. Tan, W. Liang, T. H. Lee, C. H. Choy, and Z. Shen. Design and development of a feedback mechanism and approach for patient-instrument stabilization during office-based medical procedures. In *2013 Seventh International Conference on Sensing Technology (ICST)*, pages 520–525, Dec 2013.

[Utt15] Marcus Utter. Indoor positioning using ultra-wideband technology. In *UPPSALA UNIVERSITET*, December 2015.

[WCD+16] P. Wang, C. Chen, C. Dong, H. Xu, and F. Tian. The analysis method of video camera's motion based on optical flow and slam. In *2016 International Conference on Audio, Language and Image Processing (ICALIP)*, pages 62–66, July 2016.

[Wes] Chris F. Westbury. Bayes' rule for clinicians: an introduction. In *Department of Psychology, University of Alberta, Edmonton, AB, Canada.*

[WH] & Eike Meerbach Wilhelm Huisinga. Markov processes for everybody. In *DFG Research Center Matheon, Berlin.*

[Woo07] Oliver J. Woodman. An introduction to inertial navigation. In *Technical reports published by the University of Cambridge Computer Laboratory*, August 2007.

[WRSB97] M. Z. Win, F. Ramirez-Mireles, R. A. Scholtz, and M. A. Barnes. Ultra-wide bandwidth (uwb) signal propagation for outdoor wireless communications. In *1997 IEEE 47th Vehicular Technology Conference. Technology in Motion*, volume 1, pages 251–255 vol.1, May 1997.

[WS02] M. Z. Win and R. A. Scholtz. Characterization of ultra-wide bandwidth wireless indoor channels: a communication-theoretic view. *IEEE Journal on Selected Areas in Communications*, 20(9):1613–1627, Dec 2002.

[WSFI] David M. Bevly Warren S. Flenniken IV, John H. Wall. Characterization of various imu error sources and the effect on navigation performance. In *Auburn University.*

[WYL17] K. Wen, K. Yu, and Y. Li. Nlos identification and compensation for uwb ranging based on obstruction classification. In *2017 25th European Signal Processing Conference (EUSIPCO)*, pages 2704–2708, Aug 2017.

[WZD+19] W. Wang, Z. Zeng, W. Ding, H. Yu, and H. Rose. Concept and validation of a large-scale human-machine safety system based on real-time uwb indoor localization. In *2019 IEEE/RSJ International Conference on Intelligent Robots and Systems (IROS), Macau, China)*, volume pp. 201-207, 2019.

[XTC18] Y. Xu, G. Tian, and X. Chen. Enhancing ins/uwb integrated position estimation using federated efir filtering. *IEEE Access*, 6:64461–64469, 2018.

[XWM+15] Z. Xiao, H. Wen, A. Markham, N. Trigoni, P. Blunsom, and J. Frolik. Non-line-of-sight identification and mitigation using received signal strength. *IEEE Transactions on Wireless Communications*, 14(3):1689–1702, March 2015.

[XZYN16] Jiang Xiao, Zimu Zhou, Youwen Yi, and Lionel M. Ni. A survey on wireless indoor localization from the device perspective. *ACM Comput. Surv.*, 49(2):25:1–25:31, June 2016.

[Yan18] X. Yang. Nlos mitigation for uwb localization based on sparse pseudo-input gaussian process. *IEEE Sensors Journal*, 18(10):4311–4316, May 2018.

[YB06] X. Yun and E. R. Bachmann. Design, implementation, and experimental results of a quaternion-based kalman filter for human body motion tracking. *IEEE Transactions on Robotics*, 22(6):1216–1227, Dec 2006.

[YDH16] V. Yajnanarayana, S. Dwivedi, and P. Händel. Multi detector fusion of dynamic toa estimation using kalman filter. In *2016 IEEE International Conference on Communications (ICC)*, pages 1–6, May 2016.

[YWO13] T. Ye, M. Walsh, and B. O'Flynn. Ieee 802.15.4a uwb-ir ranging with bilateral transmitter power control methodology for multipath effects mitigation. In *24th IET Irish Signals and Systems Conference (ISSC 2013)*, pages 1–6, June 2013.

[YXW17] P. Yang, J. Xu, and S. Wang. Position fingerprint localization method based on linear interpolation in robot auditory system. In *2017 Chinese Automation Congress (CAC)*, pages 2766–2771, Oct 2017.

[ZGL17] Faheem Zafari, Athanasios Gkelias, and Kin K. Leung. A survey of indoor localization systems and technologies. *CoRR*, abs/1709.01015, 2017.

[ZLW18a] Z. Zeng, S. Liu, and L. Wang. Nlos detection and mitigation for uwb/imu fusion system based on ekf and cir. In *2018 IEEE 18th International Conference on Communication Technology (ICCT)*, pages 376–381, Oct 2018.

[ZLW18b] Z. Zeng, S. Liu, and L. Wang. A novel nlos mitigation approach for tdoa based on imu measurements. In *2018 IEEE Wireless Communications and Networking Conference (WCNC)*, pages 1–6, April 2018.

[ZLW18c] Z. Zeng, S. Liu, and L. Wang. Uwb/imu integration approach with nlos identification and mitigation. In *2018 52nd Annual Conference on Information Sciences and Systems (CISS)*, pages 1–6, March 2018.

[ZLWW17] Z. Zeng, S. Liu, W. Wang, and L. Wang. Infrastructure-free indoor pedestrian tracking based on foot mounted uwb/imu sensor fusion. In *2017 11th International Conference on Signal Processing and Communication Systems (ICSPCS)*, pages 1–7, Dec 2017.

[ZWL19] Z. Zeng, L. Wang, and S. Liu. An introduction for the indoor localization systems and the position estimation algorithms. In *2019 Third World Conference on Smart Trends in Systems Security and Sustainablity (WorldS4)*, pages 64–69, 2019.

[ZYC+13] W. Zhang, Q. Yin, H. Chen, F. Gao, and N. Ansari. Distributed angle estimation for localization in wireless sensor networks. *IEEE Transactions on Wireless Communications*, 12(2):527–537, February 2013.

[ZYW+19] Z. Zeng, W. Yang, W. Wang, L. Wang, and S. Liu. Detection of the los/nlos state change based on the cir features. In *2019 Third World Conference on Smart Trends in Systems Security and Sustainablity (WorldS4)*, pages 110–114, 2019.

In der Reihe *„Forschungsberichte aus dem Lehrstuhl für Regelungssysteme“*, herausgegeben von Steven Liu, sind bisher erschienen:

1	Daniel Zirkel	Flachheitsbasierter Entwurf von Mehrgrößenregelungen am Beispiel eines Brennstoffzellensystems ISBN 978-3-8325-2549-1, 2010, 159 S.	35.00 €
2	Martin Pieschel	Frequenzselektive Aktivfilterung von Stromoberschwingungen mit einer erweiterten modellbasierten Prädiktivregelung ISBN 978-3-8325-2765-5, 2010, 160 S.	35.00 €
3	Philipp Münch	Konzeption und Entwurf integrierter Regelungen für Modulare Multilevel Umrichter ISBN 978-3-8325-2903-1, 2011, 183 S.	44.00 €
4	Jens Kroneis	Model-based trajectory tracking control of a planar parallel robot with redundancies ISBN 978-3-8325-2919-2, 2011, 279 S.	39.50 €
5	Daniel Görges	Optimal Control of Switched Systems with Application to Networked Embedded Control Systems ISBN 978-3-8325-3096-9, 2012, 201 S.	36.50 €
6	Christoph Prothmann	Ein Beitrag zur Schädigungsmodellierung von Komponenten im Nutzfahrzeug zur proaktiven Wartung ISBN 978-3-8325-3212-3, 2012, 118 S.	33.50 €
7	Guido Flohr	A contribution to model-based fault diagnosis of electro-pneumatic shift actuators in commercial vehicles ISBN 978-3-8325-3338-0, 2013, 139 S.	34.00 €

8	Jianfei Wang	Thermal Modeling and Management of Multi-Core Processors ISBN 978-3-8325-3699-2, 2014, 144 S. 35.50 €
9	Stefan Simon	Objektorientierte Methoden zum automatisierten Entwurf von modellbasierten Diagnosesystemen ISBN 978-3-8325-3940-5, 2015, 197 S. 36.50 €
10	Sven Reimann	Output-Based Control and Scheduling of Resource-Constrained Processes ISBN 978-3-8325-3980-1, 2015, 145 S. 34.50 €
11	Tim Nagel	Diagnoseorientierte Modellierung und Analyse örtlich verteilter Systeme am Beispiel des pneumatischen Leitungssystems in Nutzfahrzeugen ISBN 978-3-8325-4157-6, 2015, 306 S. 49.50 €
12	Sanad Al-Areqi	Investigation on Robust Codesign Methods for Networked Control Systems ISBN 978-3-8325-4170-5, 2015, 180 S. 36.00 €
13	Fabian Kennel	Beitrag zu iterativ lernenden modellprädiktiven Regelungen ISBN 978-3-8325-4462-1, 2017, 180 S. 37.50 €
14	Yun Wan	A Contribution to Modeling and Control of Modular Multilevel Cascaded Converter (MMCC) ISBN 978-3-8325-4690-8, 2018, 209 S. 37.00 €
15	Felix Berkel	Contributions to Event-triggered and Distributed Model Predictive Control ISBN 978-3-8325-4935-0, 2019, 185 S. 36.00 €
16	Markus Bell	Optimized Operation of Low Voltage Grids using Distributed Control ISBN 978-3-8325-4983-1, 2019, 211 S. 47.50 €

17	Hengyi Wang	Delta-connected Cascaded H-bridge Multilevel Converter as Shunt Active Power Filter	
		ISBN 978-3-8325-5015-8, 2019, 173 S.	38.00 €
18	Sebastian Caba	Energieoptimaler Betrieb gekoppelter Mehrpumpensysteme	
		ISBN 978-3-8325-5079-0, 2020, 141 S.	37.00 €
19	Alen Turnwald	Modelling and Control of an Autonomous Two-Wheeled Vehicle	
		ISBN 978-3-8325-5205-3, 2020, 175 S.	41.00 €
20	Zhuoqi Zeng	Ultra-wideband Based Indoor Localization Using Sensor Fusion and Support Vector Machine	
		ISBN 978-3-8325-5229-9, 2021, 152 S.	51.00 €